ENGINEERING GENESIS
THE ETHICS OF GENETIC ENGINEERING IN NON-HUMAN SPECIES

Working Group of the
Society, Religion and Technology Project

Church of Scotland

edited by Donald and Ann Bruce

EARTHSCAN
Earthscan Publications Ltd, London

First published in the UK in 1998 by
Earthscan Publications Limited

A catalogue record for this book is available from the British Library

ISBN: 1 85383 570 6 (paperback)
      1 85383 571 4 (hardback)

Text photographs, pages 15 and 90 © Roslin Institute

Typesetting and page design by PCS Mapping & DTP, Newcastle upon Tyne
Printed and bound by Biddles Ltd, Guildford and Kings Lynn
Cover design by Andrew Corbett
Cover photograph (top) © Murdo MacLeod
*'Dolly's eye view of the paparazzi – a social comment on
how we treat the issues?'*
Cover photograph (bottom) © Scottish Crop Research Institute

For a full list of publications please contact:

Earthscan Publications Limited
120 Pentonville Road
London N1 9JN
Tel: (0171) 278 0433
Fax: (0171) 278 1142
Email: earthinfo@earthscan.co.uk
http://www.earthscan.co.uk

Earthscan is an editorially independent subsidiary of Kogan Page Limited and
publishes in association with WWF-UK and the International Institute for
Environment and Development

This book is printed on elemental chlorine free paper

CONTENTS

LIST OF FIGURES, TABLES AND BOXES

FIGURES

TABLES

BOXES

LIST OF ACRONYMS AND ABBREVIATIONS

AAT	alpha–1–antitrypsin
ACNFP	Advisory Committee on Novel Foods and Processes
ACRE	Advisory Committee on Releases to the Environment
BBSRC	Biotechnology and Biological Sciences Research Council
BSE	bovine spongiform encephalopathy
BST	bovine somatotrophin
Bt	*Bacillus thuringiensis*
CBI	Confederation of British Industry
DDT	dichloro-diphenyl-trichloro-ethane
DETR	Department of Environment, Transport and the Regions
EPA	Environmental Protection Act
EPO	European Patent Office
FAO	Food and Agriculture Organisation (United Nations)
FDA	Food and Drug Administration (US)
GATT	General Agreement on Tariffs and Trade
GMO	genetically modified organism
HIV	human immunodeficiency virus
IVEM	Institute for Virology and Environmental Microbiology
IVF	*in vitro* fertilisation
MAFF	Ministry of Agriculture, Fisheries and Food
MEP	Member of the European Parliament
NIABY	not in anybody's back yard
NIMBY	not in my back yard
NP	nuclear polyhedrosis virus
OECD	Organisation for Economic Cooperation and Development
PCR	polymerase chain reaction
PG	polygalacturonase
PTO	Patent and Trademark Office
rBST	bovine somatotrophin artifically produced by recombinant DNA technology
RCEP	Royal Commission on Environmental Pollution
RNA	ribonucleic acid
SAGB	Senior Advisory Group on Biotechnology
SCRI	Scottish Crop Research Institute
SRT	Society, Religion and Technology Project (Church of Scotland)
TUC	Trades Union Congress
UPOV	Union for the Protection of Varieties (of Plants)

LIST OF CONTRIBUTORS

GENETIC ENGINEERING WORKING GROUP OF THE SOCIETY, RELIGION AND TECHNOLOGY PROJECT

Dr Mike Appleby Senior Lecturer in Animal Welfare, University of Edinburgh.

Professor David Atkinson Deputy Principal, Scottish Agricultural College, Edinburgh.

Mrs Ann Bruce Formerly an animal breeding specialist in the agricultural industry. (Secretary.)

Dr Donald Bruce Director of the Society, Religion and Technology (SRT) Project, Church of Scotland. (Chairman.)

Professor John Eldridge Professor of Sociology, University of Glasgow.

Reverend Dr Michael Northcott Senior Lecturer in Christian Ethics and Practical Theology, University of Edinburgh.

Professor Joyce Tait Visiting Professor, Research Centre for Social Sciences, University of Edinburgh.

Professor Ian Wilmut Principal Investigator, Roslin Institute, Edinburgh.

Professor Michael Wilson Deputy Director, Scottish Crop Research Institute, Invergowrie.

Professor Peter Wilson CBE General Secretary, Royal Society of Edinburgh.

This study is the product of a working group, and individual authors are identified only in the Case Studies in Chapter 2. Since it sets out to reflect a range of opinions, not every member of the group necessarily agrees with every point of view expressed in this book, but it is a product of the group as a whole.

ACKNOWLEDGMENTS

The Society, Religion and Technology (SRT) Project of the Church of Scotland would like to thank the participants of the working group for their hard work, openness and great sense of humour which made the discussions such enjoyable and fruitful occasions. Our sincere thanks are also due to SRT's secretary Kay Shanks for her patient and unfailing administrative help. We are indebted to Rachel Mowbray and Lisa MacDonald for their work in preparing the diagrams. The working group raises its glass (usually of water) to Kirsten Cook and her staff at the Netherbow café for many memorable meals to set us up for an evening's work. We are also very grateful to the following organisations for their time and willingness to discuss these issues: PPL Therapeutics, the Roslin Institute, the Scottish Crop Research Institute (SCRI) and Pharming Oy (Finland); for their help in lending photographs and providing diagrams: the Roslin Institute, SCRI, the Biotechnology and Biological Sciences Research Council, the International Service for the Acquisition of Agri-biotech Applications and the Scottish Agricultural College; and to the following individuals for their comments and discussion on the text: Michael Banner, Mike Bruce, Brian Heap, Dick Kolega, John Polkinghorne, Michael Reiss, and Chris Wigglesworth; and for their useful discussions: Isabel Arnal, Steven Bishop, John Bryant, Grahame Bulfield, Alan Colman, José Elizalde, Brian Forster, Louise Foster, Philippa Gannon, Harry Griffin, Tom Hartman, Bob Hay, Clive Holland, Rudolf Jaenisch, Ron James, Gudrun Kordecki, Graham Laurie, Maurice Lex, Sheila MacLean, Peter Markham, Ben Mepham, Asko Mäki-Tanila, Oliver O'Donavan, David Porteous, Julie Rankin, Egbert Schroten, Egbert Schuurman, Thomas Schweiger, Pauli Seppänen, David Shapiro, Alison Spaulding, Hein van der Steen, Sandy Thomas, Dominique Vandergheynst, Christine von Weizsäcker, Monica Winstanley and John Woolliams.

Dr Donald Bruce and Ann Bruce
Edinburgh
April 1998

INTRODUCTION

GENETIC ENGINEERING HAS ARRIVED

Only within the moment of time represented by the present
century has one species – man – acquired significant power
to alter the nature of his world
Rachel Carson, *Silent Spring*,[1] p 231

Not many years ago, genetic engineering was just one academic area among many in the biological sciences. Within a surprisingly short period of time, the ability to isolate and transfer genes within and across different species has turned into one of the biggest growth areas in scientific research the world over, and one of the hottest areas of ethical debate. New discoveries are being announced almost by the week, and a technology is rapidly emerging whose products and applications are beginning to appear in our society. The prospect of genetically engineered food on our tables, crop plants in our fields and pigs' hearts in our bodies has either become reality, or has necessitated some serious ethical thinking about what should and should not become reality. The most spectacular example of the impact of biotechnology has been the unprecedented worldwide stir over the implications of cloning, following the breakthrough in sheep nuclear transfer at the Roslin Institute in Edinburgh. But it arose out of a piece of genetic engineering whose novelty in many ways encapsulates the revolution which is taking place.

Many people suffer from the debilitating lung disease emphysema. This is caused by a genetic defect, in which the lungs do not make enough of a protein called alpha–1–antitrypsin (AAT), which regulates the amount of an enzyme in the lung wall. The result is damage to the lung wall, which can eventually be fatal. Genetic engineering could offer at least three alternative methods for treating the disease, involving humans, animals or plants. In theory, it might be possible to go to the root of the problem using human gene therapy, attempting to incorporate enough of a nondefective gene into the lungs of the patient to increase the production of the protein. Such an approach is probably a long way off, but, as an alternative, genetic modification could be used to create novel ways of producing the protein in another biological organism, on behalf of humans. It could then be purified and administered to patients as a conventional drug. Both animals and plants might offer this possibility.

In the 1980s, scientists at the Roslin Institute hit on the idea of producing AAT in the milk of sheep, by introducing a human gene into the sheep which 'codes for' this protein (that is, it sends a message which tells the body to produce it) in the mammary gland. The result was the sheep known as Tracy, and her progeny at PPL Therapeutics. Sheep produced in this way are now producing the protein for clinical trials. Roslin astonished the world in early 1997 when they announced they had cloned a sheep from the mammary gland cells of a ewe. Although Dolly became a famous celebrity overnight, she was really a sideline to the main research aim. This was to apply the technique of nuclear transfer to produce transgenic farm animals from a cell culture, something which had not hitherto been possible. This method happens also to produce cloned animals. A few months later Roslin and PPL Therapeutics produced the transgenic cloned sheep Polly from a cell line of genetically modified foetal tissue. This result may prove the precursor of many new possibilities for performing genetic modification in animals.

Meanwhile at the Scottish Crop Research Institute (SCRI) at Invergowrie, near Dundee, another breakthrough has opened up a way to produce vaccines, or therapeutic or industrial proteins, in plant tissue. This is achieved by genetically modifying a normal plant virus. The modified virus acts on the plant in such a way that it causes significant quantities of the relevant protein to be made in the leaves or other tissues. After harvesting the plant, the protein can be extracted and purified. This offers a straightforward way of producing a wide range of pharmaceutically useful proteins, and, perhaps, AAT.

Biotechnological Boom?

These examples of novel medical applications of genetic engineering represent the tip of an iceberg in a field that is growing at a bewildering rate. After the chemical revolution in the second half of the twentieth century, for many people biotechnology offers the next great hope in the quest for human security in food, resources and wealth. We are faced with an exponential growth in population, the diminution of agricultural land and other resources, and threats to the environment. Many politicians, scientists and industrialists look hopefully to genetic engineering to play a key role in feeding, resourcing and cleaning up past spillages and spoilages in the environment for a better future. In order to feed the burgeoning populations of the less developed world in the first half of the next century, some see it as inevitable that we will need to use genetic techniques to derive new drought resistant or high productivity crops, or new generations of seeds which will be naturally resistant to fungi, viruses and pests, with reduced need for spraying with artificial chemicals. Marker genes are assisting conventional selective breeding methods in animals and plants to identify and control important traits for development.

Many see genetic engineering at the cutting edge of both future resources and the environmental crisis. Genetically engineered micro-organisms hold tremendous potential in dealing with sewage, pollution and oil spills, and in providing low toxicity alternatives in chemical production. More far reaching developments include the hope that genetically engineered crops might eventually become viable substitutes for oil as carbon based fuels, chemical feedstocks and perhaps even some minerals, which would be renewable and would not exacerbate global warming.

There are also prospects of large economic and employment benefits. Biotechnology has been big business for many years but, with such potential, genetic engineering could take it into a new phase. Most developed countries are investing heavily in research, and new gene based companies are springing up to harness the new research discoveries. In Europe, the European Union (EU) sees genetic engineering as a major source of future economic growth, worth many billion ecus, and potentially providing large numbers of highly skilled, high added value jobs.[2] The EU funds major research programmes seeking to enable its member states to maintain their market position against severe competition from the US and the Pacific Rim.

Biotechnological Bust?

It is easy to wax lyrical about the potential food, health, environmental and economic benefits from these new developments, but at the same time neglect their important ethical, social and spiritual implications. The rhetoric of their potential for some areas of human life needs to be balanced against the effects on other, equally vital, aspects of human society and also on the very animal and plant kingdoms to which we are looking to provide us with this bonanza. Should we be doing these sorts of things to our fellow creatures with whom we share the planet? How do we balance the harm we might do to them against the benefits to ourselves? And to what extent are applications to animals a short step to questionable uses in humans?

Do we know as much as we think we know? The ideals of science practised in humility and with due caution can easily be pushed aside by academic success and commercial prospects. Exaggerated claims can give the impression of having the technology more in our control than we actually do. The recent history of science based technology suggests that new technological developments often suffer from myopia, and seldom turn out to be quite as straightforward or utopian as their proponents and backers suggest. There are concerns that we may be putting ourselves unnecessarily at risk in proceeding too fast after the goals we can see, and turning a blind eye to the problems we would rather not see.

As we are now seeing with the motor car, once we develop social dependencies on a new technology, it becomes very difficult to change if unforeseen problems start to emerge later. It is pertinent to ask to what extent our society has, as it were, swallowed whole certain perspectives and assumptions that technological progress is the only possible way forward, without pausing to ask if there are better alternatives. Are we indeed, as some say, 'playing God', violating something which we should not be seeking to change in the natural order? Will it open up a Pandora's box which will cause misery and oppression of the poor, instead of the promised riches and blessings? Innovation requires social approval, and a wider acceptance of the challenges to lifestyle and values which it will bring. A society should always have the right to say 'yes' or 'no'.

Something in the Air

It is also easy to spell out emotional and doom-laden prophecies, however. To raise alarm without due cause or on the basis of mere speculation can be damaging to society. For some the prospect of technological change is a matter of concern and insecurity – a threat posed by the new, the upsetting of the established patterns, the fear of the unknown, and the uneasiness over what would happen if it went wrong. For others it creates enthusiasm – seeing new potential and welcoming the change. For the latter, the concerns may be seen as the irrational fears of pessimists, or those opposed in principle to new technology, or who are not prepared to change with the times, who harbour romantic notions of alternative solutions or past ages that never were. A common attitude in scientific and official circles asserts that we should be cool and rational about these matters, and not be carried away with emotional concerns; that the public, once educated, will warmly embrace the new developments. Against this rather dismissive view, however, there is also a growing awareness that there could indeed be something real behind the misgivings which many people express, a true intuition which represents something that mere reason is oblivious to. There is also a belated appreciation that the views of the proponents of biotechnology are not so cool and rational after all, but are laden with many value judgements of their own.

Last, but by no means least, there is the spiritual dimension to be considered. Where do these developments leave us spiritually? If human life is more than material growth and scientific progress, we need to ask how the latter will affect our awareness and our relationship to God, our fellow human beings and the rest of the creation. The question of what we can do can only be truly answered having first asked what we should be.

ABOUT THIS REPORT AND THE WORKING GROUP

In December 1993, the Society, Religion and Technology (SRT) Project of the Board of National Mission of the Church of Scotland (see Appendix 4) set up a multidisciplinary working group to look at ethical issues in the genetic engineering of non-human species. Its members were chosen for their expertise in a variety of fields relevant to the issues. These included specialists in the genetics of animals, plants and micro-organisms, in animal welfare, developing world applications, issues of risk and public perception, sociology, the environment and ethics. With this spread of expertise and viewpoints, the group has examined many of the complex issues which the advances in genetic engineering are bringing to light, including a good many of the questions alluded to above. The report does not seek to follow any one theory of moral philosophy, in a logical and worked-through set of arguments. Rather, it presents the iterative process of a diverse group of people studying issues, where each has brought the insights of their different disciplines to bear.

Our aim has been to strike a balance between the extremes of optimism or pessimism in which these issues are often framed in the media, and to draw on the insights which such a varied group brings. Our membership included those enthusiastic for and those sceptical of the technology, and others who approached it undecided. In this sense, our work represents something of a microcosm of the broader societal debate, while, admittedly, comprising a set of articulate experts who are not a typical slice of society. We know of no one who would not feel they were a lay person in at least some of the diverse fields we discussed. Consequently, we have tried to write for a general, non-expert readership, providing enough technical content to allow an appreciation of the science without going into undue detail. We have sought to explain the basic concepts of both genetics and ethics, and to cut to the minimum the jargon with which both spheres abound! Having aimed at a wider audience, we hope that this book will have a special appeal for those who are more directly engaged with these fields – for scientists, for students of genetics, agriculture, environment, medicine, ethics and theology, and also for industrialists, environmental organisations and politicians.

The study has been initiated by the Church of Scotland, and while this book stems from a basic Christian theological motivation, our aim is much wider than the Christian community. This is reflected in the mixed composition and way of working of the group. Indeed, one of the most important factors of our study has been the process of discussion and debate which has been generated by the different perspectives, beliefs, disciplines and experiences of the group members. This has, we believe, provided a richness and balance to our study which makes it unique among other works in this growing field. We present this book not so much to declare our agreed positions – for we have often agreed to differ – but rather as the result of our process of learning together. None of us was an expert in more than one facet of a brilliant diamond of issues. For each it has proved a valuable and stimulating journey of understanding, stretching our minds and our hearts into unfamiliar territory, but usually with at least one of us who knew how to read the map at any given point on the trail.

It is an unfinished task, for several reasons. Firstly, we could not possibly cover all the ground we would have liked, and have had to be selective in our explorations. There is much uncharted territory, and probably one or two dragons we have missed. Secondly, the technology is moving so fast that we are only too aware that by the time this text is read, many aspects will have moved on. Two vital issues emerged at a relatively late stage in our drafting, the explosion of interest in mammalian cloning and the importation into Europe of genetically modified soya and maize. Tomorrow's developments may well raise new issues, and some of the old ones may in time seem less relevant. But today, we do not have the benefit of this hindsight!

The study is also unfinished in a third and perhaps most important way. We have not written the last word on any of these issues. We present our work to offer guidance and insight to those puzzled by the complexity of the issues, and who are seeking some stimulating and well informed views as well as some direction for how to look at these issues for themselves. But we also present them as an continuing process, as an invitation for you, the reader, to join in an ongoing dialogue with us. Your thoughts may add to the continuing reflection process. So we invite you to reflect with us, and, if you will, to share your reactions and thoughts with us, as we have sought to share ours with you.

1 EXPLAINING GENETIC ENGINEERING AND ITS USES

SHAPING THE CREATION

In the book of Genesis, the ancient biblical account of beginnings, God invites human beings not simply to flourish and reproduce their own kind, but also to take control over and care for the Earth as one might work a garden, to name the animals, to search for minerals, to relax and enjoy 'the cool of the evening', and to do it all in a warm and reciprocal relationship with God.

Since flint axes, knives and arrowheads, humans have always practised technology. From the earliest times, they have adapted the natural world to serve their own ends. This is a feature of human living. At the most basic level, humans have found ways to channel and shape natural things and processes to meet the necessities of providing themselves with food, water, shelter, warmth, clothing and so on. This has included domesticating and using living creatures, both plants and animals. Having achieved a measure of security, our energies need not be exhausted just in ensuring day-to-day survival. A myriad of other possibilities open up before us – to inquire, to create culture, art and civilisation, and to do new things with the natural environment. As we learn to innovate, and especially as modern science leads us to an unprecedented understanding of the workings of the natural order, new devices, ideas and methods are discovered.

Technology always comes out of some social context. A new invention may have started off to solve a recognised public need, with general approval. In our own times science has become a complex specialist discipline, remote from ordinary people. It is now far more likely that the scientists surprise society with what they have discovered. Again, those with political or socioeconomic power may have commissioned the new development and then simply imposed it upon others. In whatever way a technology is assimilated, however, it in turn remoulds society in some way. Things are never quite the same afterwards. It shapes our horizons and expectations. Western society in particular has come to expect technological advances and a continuous, incremental rise in living standards, security and wealth.

Technology thus brings challenges to society. Often changes are only minor, but from time to time a new development poses radical questions to our assumptions and self-understanding, to lifestyles and habits, to social patterns, and to our ways of relating to each other and to the natural order. Genetic engineering is one such development. The ability to manipulate the biological world at its most fundamental levels – the heritable genetic code of the deoxyribonucleic acid (DNA) molecule and processes of the cell nucleus – ranks as one of the most important developments in technology of the twentieth century. Although in some ways it can be seen as part of a chain of developments from the technology of the past, it is a way of shaping creation that has begun to raise some of the most profound questions of our times.

In the first instance, it poses these questions for developed Western societies, like Scotland, whose fabric is reasonably stable and secure, and where people have the luxury of time to ponder such issues. But the implications and effects of genetic engineering are global. If survival depends on the next crop or the next meal or a drinkable water supply, then the ethics of applying new technology assumes a lesser significance.

It is in this context that this chapter sets out to explain something of what genetic engineering is, for those unfamiliar with it, and to illustrate something of the range of applications of gene technology to animals, plants and micro-organisms, by way of introduction to the 11 case studies. The theme of this book is how this world interacts with the world of ethics and values, whose concepts and terminology can seem equally unfamiliar to many scientists. Some of the basic tools and terms used in ethics will be explained in Chapter 3, as well as how these are applied to the fundamental issues which underlie genetic engineering. Appendix 3 will suggest some methods which readers may find helpful to assess the ethical issues for themselves.

AN INTRODUCTION TO GENETIC ENGINEERING CONCEPTS AND TECHNIQUES

What is Genetic Engineering?

Genetic engineering is a very broad term which covers a range of ways of manipulating the genetic material of an organism. It is also variously called gene manipulation, genetic manipulation, recombinant DNA technology, the new genetics, targeted genetics and, in humans only, gene therapy. In popular thinking it is frequently confused with cloning, which is not at all the same. The cells of living organisms contain genetic material which regulates the processes of the organism. This genetic material consists mostly of the complex chemical known as DNA, although sometimes it involves the related chemical RNA (ribonucleic acid). Pieces of this genetic material form genes and it is the ability to identify and manipulate one or more of these genes which underlies genetic engineering. It is estimated that there are something like 100,000 genes in a mammal and about 80,000 in a plant. Genetic engineering can involve manipulating genes both within species or between species. The products are generally referred to as genetically modified organisms (GMOs), or transgenic organisms. Since the early 1970s, genetic engineering has developed very rapidly as a powerful new tool for both the biological research sciences and the biotechnology industries, and an increasing number of applications are being brought to market.

Selective Breeding

Human beings have in a sense been performing a type of genetic engineering in the form of selective or 'classical' breeding ever since we ceased being hunter-gatherers and settled down to domesticate animals and crop plants some 10,000 years ago. Breeding involves the enhancement of recognisable traits in an organism by selecting for reproduction the individuals or populations which best exhibit the desired trait. This occurs in micro-organisms such as bacteria and the yeasts used in brewing and baking as well as in plants and animals. Some selection has taken place in response to social whims and fashions such as the desire for particular colours in domestic animals and ornamental plants. In a production context however, 'improved' varieties and breeds are selected empirically to enhance key characteristics such as higher yield, better industrial performance (eg wheat for baking, barley for making whisky) or greater disease resistance.

The process of selective breeding requires a number of generations of the organism. Because of the nature of growing seasons or an animal's generation cycle time this means that the process is cyclical and usually slow. Many of the traits of interest are controlled by multiple genes, each of which has a small effect on the trait. The trait may also be affected by the environment in which the organism finds itself. It may become difficult to discern whether an animal has the desirable characteristic because of its genetic inheritance or because of the environment it has experienced. Making the necessary breeding judgements can be highly sophisticated, using complex statistical analyses of the different traits, but it still ultimately relies on the outward appearance or the measurements of traits to indicate the genetic value of the organism.

The above factors mean that selective breeding is by nature somewhat imprecise and unpredictable. Another important source of imprecision arises because natural sexual reproduction involves exchanging and mixing thousands of genes from each partner. The resulting offspring inherit an *almost* random mixture of the characteristics of both parents. The mixing is not completely random, hence some characteristics tend to be inherited together. When a breeder selects a crop plant or a cow for one particularly desirable trait, such as larger seeds or more milk, huge numbers of other genes are also selected. These genes confer a range of other traits, many of which are unknown or biologically neutral. Sometimes, however, genes for some undesirable characteristics such as predisposition to a disease may also be present in the selected progeny. A kind of halfway stage between classical breeding and genetic engineering is of increasing interest to animal breeders. This is called marker-assisted selection, and involves using genetic markers which are associated with one of the desired traits in order to identify the 'best' animals or plants to produce the next generation. Other genetic changes can be induced using chemicals or irradiation, as discussed in Case Study 2.

In contrast to classical breeding, genetic engineering now offers the capacity to add a single gene, or a small cluster of genes which control a new trait. This would be added to an already tried and tested variety of a given species. In some cases genes can also be deleted or have their function disabled. One major difference between this and classical breeding is that genetic engineering does not involve mixing the entire complement of genes between two individuals, but only the particular genes which have been identified. Genetic engineering should therefore be more rapid, precise and predictable when compared with selective breeding methods. The latter have jokingly, but not entirely unfairly, been summarized as: 'pick and cross two of the best, then hope for the best!' Genetic engineering is, however, still a relatively young science, and, for all its successes, it is not without its hit and miss elements.

The Principles of Genetic Manipulation

DNA

The fundamental chemical and biological discoveries which made genetic manipulation possible can be traced back to early studies on bacterial

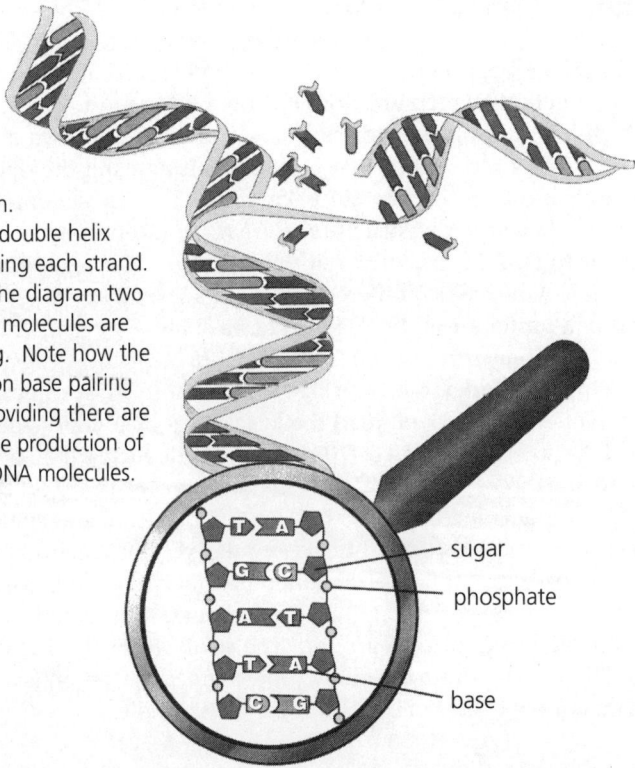

DNA replication.
A parent DNA double helix
unwinds exposing each strand.
At the top of the diagram two
daughter DNA molecules are
shown forming. Note how the
process relies on base pairing
and results, providing there are
no errors, in the production of
two identical DNA molecules.

sugar

phosphate

base

Source: Straughan and Reiss (1996)[4] reproduced by kind permission of BBSRC

Figure 1.1 *DNA Replication*

genetics in the 1940s, and to the discovery of the structure, and hence the modus operandi, of DNA (Figure 1.1) in 1953.[3]

DNA is a biological polymer which contains all the essential chemical and developmental information which the cell needs to perform the biochemical processes of life. It does this by means of a complex chemical code, which consists of different combinations of a very large number of smaller chemical groups called bases, which are attached to an inert backbone forming the helical structure of the DNA molecule. There are only four types of base, which are abbreviated by the letters A, C, G and T. It is the sequence of these bases, in different orders and combinations along the backbone, which makes up the genetic information – the genetic code. Broadly speaking, each group of three bases forms the code for a single amino acid. A chain of these amino acids will lead to the production of a protein which will affect the biochemical processes in the organism – just as letters form into words which convey to us particular meanings. Some of these bases act as punctuation marks, for example to identify the beginning and end of a protein. Other sequences of bases have functions which are not yet fully understood. The length of DNA which is required to produce a single protein is known as a gene. The gene is said to 'code for' that particular protein and the operation of the gene to produce the protein is known as the 'expression' of the gene. Further details of the structure of DNA are given in Appendix 2. RNA is similar to DNA in structure, but the chemistry of a sugar in the backbone and one of the bases is different. RNA plays several key roles in the production of a protein from the DNA code. In some organisms, such as viruses, RNA is the only form of genetic material present.

Gregor Mendel was a German monk whose study of pea plants in the nineteenth century identified the basic rules of genetics and inheritance. Work between 1940 and 1970 on the genetic material of common bacteria such as *Salmonella typhimurium* and *Escherichia coli* then provided a far more detailed operational definition of a gene and an understanding of how its expression could be regulated. Many elegant studies used mutant bacterial strains or exploited natural viruses of bacteria (known as bacteriophages) which can jump in and out of the bacterial chromosomes and move pieces of DNA around. The small size of their DNA, their rapid growth and the short time between each generation of bacteria were key factors in the effectiveness of these genetic studies.

Cutting, Pasting and Copying DNA

Genetic engineering only became feasible when a number of key discoveries had been made. These enabled DNA molecules first to be isolated from the originating organism, then 'cut and pasted' in defined ways, and introduced and integrated into the normal DNA of a recipient organism. Until the early 1970s, the only way to isolate a small piece of DNA was to shear the entire DNA from the chromosomes of a cell by mechanical force. Unfortunately, this only produced random fragments of DNA so that no two fragments were likely to contain the same sequence. The breakthrough came with the discovery of a family of enzymes which are part of a natural defence mechanism of bacteria and which are able to cut DNA at specific places. These enzymes are called restriction endonucleases and each of them recognises one particular short sequence of bases in the DNA and makes a precise cut at that point. One enzyme might recognise the sequence GAATTC, another AGTACT, for example. When any DNA is cut it always leaves the same pattern of bases where the cut was made,

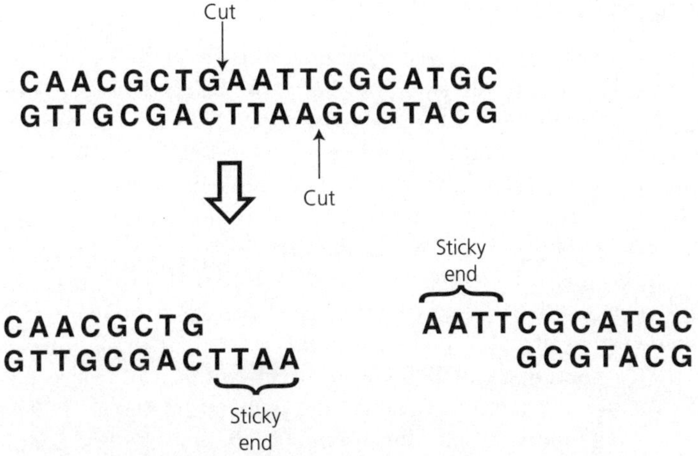

Cut

C A A C G C T G A A T T C G C A T G C
G T T G C G A C T T A A G C G T A C G

Cut

Sticky
end

C A A C G C T G   A A T T C G C A T G C
G T T G C G A C T T A A   G C G T A C G

Sticky
end

The action of the restriction endonuclease EcoRI. This finds the DNA sequence
GAATTC and then cuts the DNA between the G and A bases.

Source: Straughan and Reiss (1996)[5] reproduced by kind permission of BBSRC

Figure 1.2 *Cutting, Pasting and Copying DNA*

regardless of the organism or species. Thus, if DNA from one organism is
cut by a particular restriction endonuclease, and the same enzyme is used
to cut a fragment of DNA from another organism, the two fragments could
be stuck together at that point (Figure 1.2). In this way, particular
fragments of DNA from different sources can be made to link together.
Further details of this process are described in Appendix 2. The product
of this is called a recombinant genetic sequence, or recombinant DNA,
and it is then multiplied many times over and introduced into bacteria,
plants or animals.

The multiplication process is usually called cloning. One method of
cloning involves the repeated copying of the base sequence of the recombinant DNA by means of a carrier, which is known as a vector. The vector
can be a chromosome, plasmid or virus. A plasmid is a small closed loop
of DNA which multiplies to very high numbers in a single bacterial cell.
The new gene is attached to the vector, and when the vector goes about
its normal business of replicating itself, the new gene is automatically
copied too; see step 3 in Figure 1.3. Alternatively, a chemical method is
used, known as the polymerase chain reaction (PCR). This uses an enzyme
to make repeated copies of a section of DNA in a chemical solution.

Gene Sequencing

These techniques of cutting, pasting and cloning small, defined pieces of DNA enabled new and otherwise unobtainable combinations of one or more genes to be produced from any organism. They also permitted the rapid development of techniques for sequencing DNA – that is, to determine the sequence of bases in a given DNA molecule. Over the past 20 years, these methods have become routine and automated, and are capable of operation by robots in the large scale sequencing programmes like the human genome project and analogous projects on such organisms as yeast, rice and pigs. In their simplest form, the materials for manual DNA sequencing are available in convenient 'kit' form. In a single tube, with sub-microgram amounts of DNA in a few microlitres (a tiny drop) of reaction mixture, it is possible to obtain the sequence of 500 to 1000 bases of DNA in a few hours. To put this in context, in 1965 the first complete gene sequence was determined. This was of a single-stranded RNA of only about 80 bases from yeast. It took many years of work, many hundreds of research staff, several grams of pure RNA and won a Nobel prize!

Over the past 25 years, the techniques of cutting, pasting, cloning and sequencing DNA, of designing new gene sequences, and of making specific alterations to existing gene sequences, have all become routine tools in research and development in the biological sciences. Their application has unravelled details of complex structures and function within cells, tissues and whole organisms. It has greatly advanced our knowledge of microbial, plant and animal life both in their normal processes and in what happens during disease.

Transferring Genes

Techniques to transfer DNA have developed over the past 50 years. They began by exploiting knowledge of natural processes in bacteria where DNA molecules move between cells – the bacterial equivalent of mating. Viewed as a whole, DNA transfer methods today can be classified as either biological or mechanical. The former still rely largely on bacterial or viral DNA vectors, whereas the latter include all manner of devices from injections, guns and microbullets, to darts, sparks, abrasives, water and salt shocks!

In order to make the 'foreign' gene express itself in the new DNA background into which it has been introduced, a special sequence called a promoter is always added to the introduced gene. Promoters act as signals to determine when and where the gene is 'switched on' to produce the new protein. Plant gene promoters respond to external signals such as light/dark, drought stress, salinity, plant hormones, wounding stress, or just by being part of a cell destined to become a root not a flower, or *vice versa*. In animals, promoters can respond to particular items in the diet and can determine where the foreign protein is produced. In Case Study 8 the gene which results in the production of human alpha–1–antitrypsin in sheep was linked to a promoter which caused the protein to be produced only in the mammary gland, whereas the sheep version of the protein is produced mainly in the liver (just as the human version is in humans).

Current Gene Transfer Technology for Micro-organisms

Genetically modified bacteria are in some cases produced in their own right, as in Case Study 1 or for example in pollution control applications. Their most frequent use is, however, as the means by which other genes are multiplied. During the routine cloning of any gene, even one destined for insertion into a plant or animal, it is most likely that the early steps involved multiplication of the DNA sequence in a bacterial plasmid. Once the DNA fragments have been joined together in a test tube, they are introduced into bacterial cells which have been rendered 'competent' to take up recombinant DNA by following a carefully controlled series of growth and salt/heat/cold shock treatments. These procedures appear empirical but necessary for efficient DNA entry. Special bacterial strains are available for these procedures. These strains require selective nutrients and growth media as they have been genetically debilitated and are incapable of surviving if they escaped into the 'wild' by accident. They are also unable to recognize the incoming DNA as foreign. To select only those bacterial cells which have taken up the foreign DNA, it is usual to also include an extra gene which confers resistance to a particular antibiotic. This antibiotic is then added to the growth medium, and the only bacteria which continue to grow are those which received the recombinant plasmid molecule attached to the antibiotic resistance gene. Other techniques for uptake of engineered DNA are described in Appendix 2.

Gene Transfer Technology for Plants

Gene transfer technology for plants involves methods to introduce foreign genes into the plant cells and methods to regenerate whole plants from the transgenic plant cells. Figure 1.3 illustrates the various steps in plant genetic engineering.

Methods of Introducing Foreign Genes into Plants

Experiments on the transformation of plant cells began in 1980 and the first genetically engineered plant (tobacco) was reported in the scientific journal *Nature* in 1983. The first successful method for DNA transfer into plant cells used a natural soil-inhabiting bacterium called *Agrobacterium tumefaciens* which infects wound sites on plant stems. In the wild, *A. tumefaciens* induces cankers or crown galls by producing plant hormones which promote uncontrolled cell growth. It does this by inserting a portion of its DNA into the plant. By adding the 'foreign' gene to *A. tumefaciens* it can 'hitch a ride' into the plant cells. Further details of the process are given in Appendix 2. This natural, and so far unique, DNA transfer mechanism is efficient. It was used by scientists in Belgium, Germany, Holland, the US and the UK in the early 1980s to transfer designer genes and antibiotic resistance marker genes into plant cells in culture, into the damaged cells at cut leaf surfaces, or into stem cuttings (see step 5 in Figure 1.3). The system was primarily appropriate for broad-leafed plants. For many years, however, there were technical problems in transferring foreign DNA into cereals, bulbs, legumes or grasses, which were either not normal hosts for *A. tumefaciens*, or which presented tissue culture and plant regeneration barriers.

To transfer and integrate foreign DNA to the cells of plants outside the host range of *A. tumefaciens*, mechanical means were invented. The most theatrical and now most widely used of these was the so-called biolistic method, using a particle gun, first made in 1987 at Cornell University in the US. The genetically engineered DNA is precipitated onto the surface of extremely small gold or tungsten particles (less than a millionth of a metre across). The particles are then placed on the blunt front end of a macro-projectile which is fired at a stopping plate with a small central hole using a blank cartridge or high pressure burst of helium gas. The macro-projectile hits the plate, but the gold or tungsten micro-projectiles keep moving and hit the plant leaf, root, stem, cell suspension, callus or whatever tissue is placed under the stopping plate. Those cells which get bombarded too violently die, while those on the outer perimeter of the blast do not receive enough DNA-coated gold or tungsten to become transformed. However, a ring of cells midway receives enough DNA at the correct speed and force to penetrate the thick plant cell walls, the delicate cell membranes and the nuclear membrane to enter the nucleus of the cell which contains the DNA. Once inside the nucleus the foreign DNA is somehow inserted into the plant chromosomal DNA. Numerous variations on the biolistic DNA trans-formation process have been tried and tested over the past ten years. Other less commonly used methods of gene transfer are described in Appendix 2.

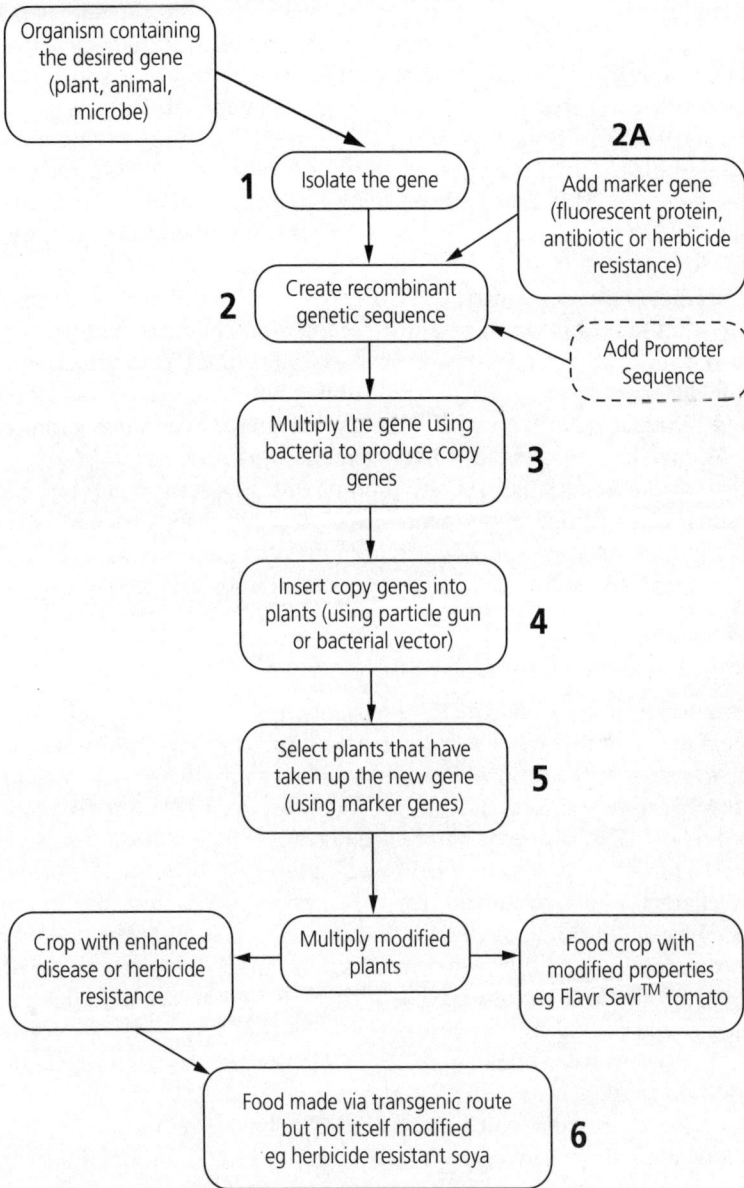

Figure 1.3 *Steps Involved in Plant Genetic Engineering*

The number of copies of foreign DNA inserted and the sites where this happens in the plant chromosomes are entirely random for a given cell. The foreign DNA may be inserted at several different sites and in several different chromosomes, but a single copy inserted at a single site is the most typical. Sometimes the DNA transfer and random integration processes place the foreign gene next to or even inside an essential plant gene sequence. This causes those plants to either to fail to grow, or to grow with some undesirable trait. Such abnormal lines are routinely discarded.

Attempts have been made to transfer more than one new gene into a plant at a time. This is important if complex biochemical pathways are required, or if multiple disease resistance is essential for a valuable but highly susceptible crop. In general, only single gene traits have been added to transgenic plants to date, but a number of ingenious strategies are now making it increasingly likely that multi-trait transgenic plants will be feasible. Methods which involve multiple independent promoters, each of which drives a single gene in the foreign DNA, have largely been unsuccessful, unless each promoter is chemically unique and the relative levels of expression of the different genes is not critical for a trait.

Tissue Culture and Regeneration

No matter how efficient the DNA transfer process, only a very small percentage of all treated cells will receive an intact foreign DNA construct at just the right time and place, and without too much cell damage to allow repair to take place. In addition, the introduced DNA must integrate into the host DNA at one or more chromosome positions where it can be expressed from its promoter. Provided all these events occur, the observable characteristic conferred by the foreign DNA should then be detectable. The problem is then to find these few cells and separate them from the millions of other cells where any or all of these events did not happen properly. To avoid having to look for these few needles in a haystack, ways are found to discard or kill selectively the unsuccessful cells. To do this, a selectable marker gene is attached to the foreign DNA in addition to the gene for the actual new trait wanted. Selectable marker genes encode proteins which make any cell which expresses them resistant to some lethal chemical such as a herbicide, an antibiotic or a toxin. Thus the 99.999 per cent of cells that were exposed to the particle gun or to *A. tumefaciens* but which failed to become transformed will die when, for example, the antibiotic is applied to the whole population. Only the survivors are likely to have the foreign DNA inserted, and can be selected for further study (see step 5 in Figure 1.3).

If kept sterile against fungal or bacterial attack and disease, each plant cell can, with careful culture conditions, become a full grown plant again. Cuttings can then be used to propagate the plant vegetatively or it can be self pollinated for seed. Each daughter plant made in this way is genetically identical to the original. Further details of this procedure are given in Appendix 2.

Once the DNA has become part of a plant chromosome it is multiplied and inherited for many generations and it is chemically indistinguishable from normal plant cell DNA. As each transgenic plant will have been regenerated from a single transformed cell, every cell in the plant will have inherited the same foreign DNA. This means that both the pollen and the ovules (and daughter seed) will carry the foreign gene(s). This potential for genes to be released into the environment explains the underlying rationale for the application of precautionary environmental risk assessment procedures (see Chapter 6).

Using Viruses, Bacteria and Fungi with Higher Plants

Sometimes it is more desirable, more feasible or more economic to genetically engineer the DNA or RNA of lower organisms such as viruses, bacteria and fungi. When these infect a plant they will then act as a transient genetic expression system within the host plant. An example of this is given in Case Study 4 with the production of vaccines from genetically engineered plant viruses. The reasons for taking this approach vary case-by-case. Often it is technically difficult to modify the genetic material of the plant itself, or it may be impossible or undesirable to do so. Sometimes, many more copies of the genetic material of the lower organism exist within each cell of the higher species than copies of its own genetic material, so that a higher yield of the desired protein can be obtained by modifying the lower organism DNA rather than the plant DNA. Similarly, sometimes the expression level of the lower species DNA (or RNA) is greater, because of the need to compete with the higher host cell, so that a higher output of the desired product can be obtained by modifying the plant virus rather than modifying the plant itself.

Gene Transfer Technology in Animals

Injecting Genes Into the Embryo

Gene transfer in animals was first achieved by the direct injection of several hundred copies of the recombinant gene into a nucleus of an early embryo. Embryos may be obtained either by fertilisation in the laboratory, using techniques that are similar to those for human *in vitro* fertilisation (IVF), or fertilised eggs can be collected from donor animals during surgery. Hormone treatment is used to increase the number of eggs that are shed before mating or artificial insemination. The abdominal cavity is opened, the reproductive tract exposed and sterile fluid is passed through the tract in order to flush out the fertilised eggs into collecting dishes. Techniques of anaesthesia and surgery are again very similar to those used in human treatments. Whereas sheep and pig embryos have been recovered during surgery, the great cost of the technique in cattle has effectively prevented this approach. Instead, cattle embryos have been obtained from unfertilised eggs in the ovaries of slaughtered cattle. These eggs are matured in the laboratory before fertilisation and culture to the stage at which they are injected. As culture techniques improve, it is increasingly likely that embryos will be produced in the laboratory using IVF techniques, but surgical recovery of fertilised eggs will always offer the advantage of having embryos from specific, selected parents.

Injecting the DNA requires the use of microscopes because mammalian embryos at this stage are only approximately 0.1 mm in diameter and can only just be seen by the naked eye. To allow the injection, fertilised eggs are held by a small suction pipette before a very fine needle is introduced into a nucleus and the liquid containing the DNA is injected until the nucleus can be seen to swell (see Figure 1.4). A proportion of the injected eggs burst within a short period and are discarded. The surviving eggs are transferred to surrogate mothers for development to term. Although the results are variable, only approximately 10 to 20 per cent of the eggs develop to become live young, and only about 1 per cent of injected eggs develop to be born as offspring carrying an additional gene. The only way of knowing which offspring carry an additional gene is by examining DNA from each animal and this is usually obtained from a blood sample soon after birth.

In the majority of cases the added gene will be transmitted to some of the offspring of the founder. As the gene is normally only integrated into one of a pair of chromosomes it will only be transmitted to half of the offspring, and experience shows that in some cases an even smaller proportion of the offspring of the founder generation inherit the gene, apparently because the gene was not present in all of the cells from which the germ cells are derived. However, in subsequent generations the gene is usually inherited by half the offspring and it is possible to use conventional breeding to produce as many animals as required.

Gene transfer by direct injection has several limitations. It is expensive because the efficiency is low and large numbers of animals are

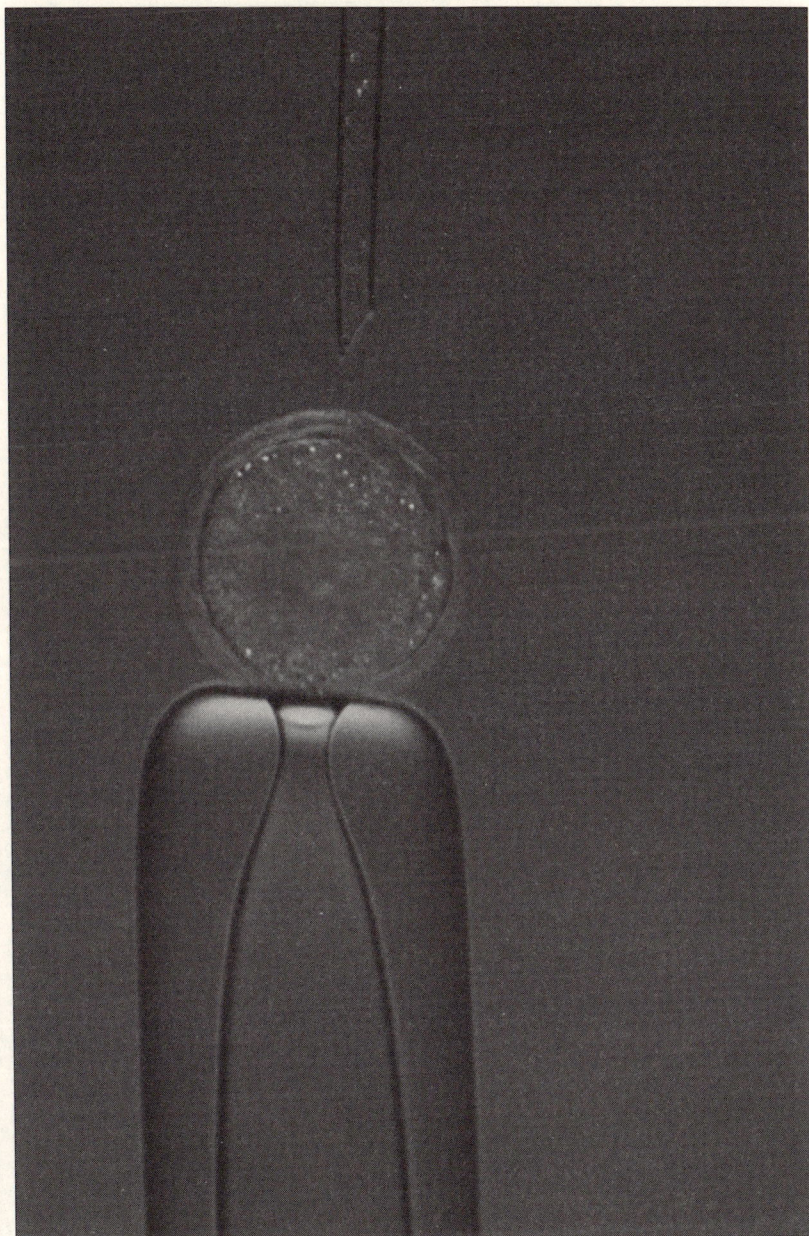

Photograph: Roslin Institute, reproduced by permission

Figure 1.4 *Micro-injection of DNA into a Cell Nucleus*

required, both as embryo donors and as surrogate mothers. It is believed that integration of the additional gene into a chromosome occurs because the act of injection causes breaks in chromosomes and that the repair mechanisms inadvertently include some of the injected DNA in the chromosome when repairing a break. This interpretation certainly accounts for the facts that the site of gene integration is apparently random and that in approximately 7 per cent of cases the gene has integrated within an endogenous gene, disrupting the function of that gene. It is also found that the functioning of transferred genes varies widely and it is believed that the neighbouring DNA at the site of integration influences the transgene. Finally, direct injection is only able to add a single gene. Despite these limitations direct injection has been used very widely in research and for a small number of specific applications, including the production of pharmaceutical proteins in the milk of farm animals (Case Study 8). The technique of injection has also been used to produce transgenic fish and some poultry. The technique is much more difficult to apply to poultry because access to the embryo at the appropriate stage of development is difficult. When an egg is laid, the embryo is too advanced in its development to allow injection to be carried out. Genetic engineering in poultry has therefore required the development of new techniques such as removing the fertilised eggs, injecting them and then allowing them to develop in an artificial shell. Because of these problems, genetic engineering has not, to date, been used extensively in poultry.

Modifying Cultured Embryo Cells

In plants, almost any individual cell can be genetically modified and then a whole new plant grown from this cell. This is not generally possible with animal cells. However, a technique has been established in mice for developing a fully grown animal from an initial cell. This technique has been used in developing many of the later examples of mouse models of human diseases referred to in Case Study 10. The technique depends upon the availability of 'embryonic stem cells' isolated from a culture of mouse embryonic cells in circumstances that allow division, but not differentiation. If such cells are introduced into another recipient embryo they sometimes retain the ability to colonise all of the tissues of the developing offspring, including the germline. The resulting offspring are derived from the cells of two embryos and are known as 'chimaeras' because some cells in the animal will have the genetic composition of the recipient and some will have the genetic composition of the donor. Offspring from some of the chimaeras, where the germline cells are from the donor, will have all their cells with the genetic composition of the donor.

While these embryonic stem cells are in culture, molecular techniques can be used to introduce precise genetic changes and in this way the change is introduced into the germline. These changes can involve deletion and modification of genes and not just addition as is possible with the micro-injection technique. It has been predicted that similar systems will probably be established in the future in other animals but, at the time of writing, embryo stem cells are not available for other species.

Transferring Nuclei Between Cells

The recently developed technique of nuclear transfer has the potential to allow the introduction of specific changes to livestock species equivalent to those possible for mice using embryo stem cells. Nuclear transfer is considered in more detail in Case Study 11 and only a brief description is given here. Nuclear transfer involves two cell types: an unfertilised egg and a donor cell (Figure 1.5). The genetic material is removed from the unfertilised egg and the genetic material of the donor cell is introduced into it, while cell development is temporarily suspended. An electric current is used to fuse donor nucleus and recipient cytoplasm and resume cell development. This has now enabled live sheep to be grown from cell cultures, achieving an end analogous to mouse stem cell technology, but by a quite different means. In each case the sheep is a clone of the animal which donated the nucleus. This was performed first with embryo cells,[6] then udder cells,[41] and then genetically modified foetal cells which produced transgenic cloned sheep.[7]

It remains to be seen how widely this can be used and in which other animals. It has now been successfully applied to cattle and mice, but only in certain cells and not with others. While the method is still at an early stage of understanding, it opens up the possibility of performing much more precise genetic modifications in farm animals at the cell culture stage, before the nuclei are used as nuclear donors. Existing methods could be used to select just the populations of cells which have incorporated the correct modification for nuclear transfer and then growth into a full sized animal. This method should enable a range of animal genetic modifications that has so far only been possible in mice. Genes could not only be added, but also removed or replaced. For example, thus far in sheep it has proved sufficient to add the gene and engineer the construct so that it is expressed in the relevant organ while the equivalent sheep gene remains active, but there are potential applications in which it would be important to switch off the animal's gene so that the human gene can perform the desired function. The new method would enable this to be done. The extension of nuclear transfer to mice also opens a wide range of medical research applications.

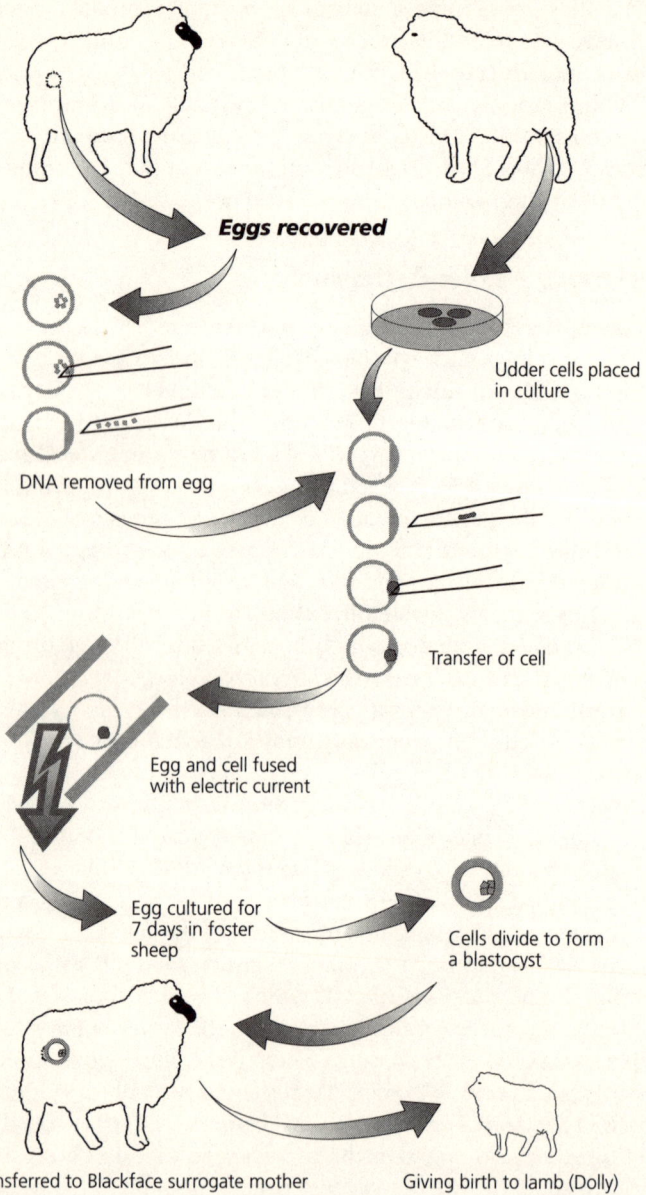

Eggs recovered

DNA removed from egg

Udder cells placed
in culture

Transfer of cell

Egg and cell fused
with electric current

Egg cultured for
7 days in foster
sheep

Cells divide to form
a blastocyst

Egg transferred to Blackface surrogate mother

Giving birth to lamb (Dolly)

Source: Roslin Institute, reproduced by permission

Figure 1.5 *Nuclear Transfer Process, as Used to Produce Dolly*

Future Targets for Technical Advances

The above description of the methods used to perform genetic modification has indicated that the science is still relatively young, incomplete and subject to rapid change. Certain areas are already quite well developed, while others are at a much more rudimentary stage. This makes it especially difficult to predict the directions which future advances will take. As with all scientific knowledge, there is no finite end point to the understanding or implementation. Ideally, each new discovery brings greater understanding, but often it also reveals more complexity, with new avenues to pursue and new hypotheses to test. It is expected that overall this will lead to an increase in our basic knowledge of the physical, chemical and cell biological events which surround the uptake and integration of foreign DNA, and of how a cell regulates its own gene expression. With such knowledge, it is reasonable to expect that many procedures in use today could be made more efficient and predictable, and that some features which raise risk issues will be removed. Some of the current target areas in which technical improvements are desired – some of which are already foreseeable, others which are more long term aims – include:

- to develop more efficient methods to transfer DNA into plant and animal cells;
- to make the way this foreign DNA is then integrated into the genome of the organism more predictable and controllable than the present rather random situation;
- to enable gene deletion and replacement in farm animals using nuclear transfer methods;
- to develop ways to select transgenic material from unmodified material non-destructively;
- to design less controversial selectable marker genes for use in plants than the current methods which rely on antibiotic or herbicide resistance;
- to compile a tool-box of promoters to control how the newly introduced gene is expressed – these may be cell-, tissue-, organ-, stimulus-, time- or quantity-specific;
- to develop reliable methods and DNA construct strategies to enable multi-gene traits, as opposed to just single genes, to be transferred to crop plants or animals;
- to develop ways to prevent foreign DNA sequences being mobilised by pollen or by other germline cell types;

- to improve the efficiency of tissue culture regeneration of the more difficult species, such as cereals, grasses, legumes, woody perennials and marine and freshwater algae;
- to design more effective and broad-range pest and disease resistance genes.

In parallel, there remains much work to be done in disciplines to which genetic advances must continue to relate, such as soil science and animal physiology and embryology. The aim is for more efficient and effective end products, with no unpredicted side effects, and a consistency of performance which would merit greater confidence of society in this area of technology.

POTENTIAL APPLICATIONS OF GENETIC ENGINEERING

As a young technology in a phase of very rapid growth and development, the range of applications of genetic modification is theoretically enormous. In practice, however, there are at present some important limitations on the types of modification which can be done, and which species they can be applied to. The following summary gives four general areas where genetic engineering is currently being applied.

Food Production Applications

One of the biggest hopes from proponents of genetic engineering is that it may enable us to feed more people from the available land, in the face of the expansion of the world's population. This may be a crucial factor in helping to preserve and secure life and dignity for people, both present and future. If it is possible to develop genetically modified plants to grow on marginal land, including drought-prone, saline, eroded or desertified land, this would not only extend food production in vulnerable areas of the world, but would also help to reverse the loss of arable land through soil erosion and desertification. Table 1.1 shows the number of genetically modified foods cleared or under consideration in the UK.

In food production management, genetic engineering might make possible higher yields of milk, eggs and meat from a given number of cows, chickens or pigs. This could reduce pressures on land use, and also lower the production of methane, ammonia and other wastes from intensively reared animals. It may allow foodstuffs to be more resistant to the

Table 1.1 *Genetically Modified Foods Considered in the UK by the Advisory Committee for Novel Foods and Processes up to March 1998*[8]

Foods cleared by ACNFP	Date cleared
GM baker's yeast	March 1990
GM brewer's yeast	February 1994
GM soya (Glyphosate resistant)*	February 1995
Oil from GM oilseed rape (fertility restorer line, male sterile line)	February 1995
Paste from GM tomato*	February 1995
Oil from GM glufosinate-ammonium tolerant oilseed rape	May 1995
Oil from GM oilseed rape (2nd fertility restorer line)	September 1995
Oil from Glyphosate-tolerant GM oilseed rape	February 1996
Flavr Savr™ tomato (GM tomato to be eaten fresh)	February 1996
Paste from GM tomato (extension to the 1995 clearance)	February 1996
Insect-resistant GM maize-processed food products	May 1996
Ribflavin from GM *Bacillus subtilis*	January 1997
Glufosinate-tolerant GM maize	February 1997
Insect-resistant GM maize	February 1997
Herbicide-tolerant GM maize	February 1997
Herbicide-tolerant GM cottonseed	February 1997
Herbicide-tolerant and insect-resistant GM maize	February 1997

Foods still under Consideration at March 1997

Processed GM tomato
Insect-resistant GM maize
Oil from GM herbicide-tolerant oilseed rape
Insect-resistant GM cottonseed
GM chicory (*Radicchio rosso*)
GM oilseed rape with high lauric acid content

Notes: * products on sale in UK

vagaries of climate, mechanical picking and transportation. It could also be used to enhance the quality of food for the consumer and help reduce wastage of perishable foods.

Improved Yield from Crops and Animals

Of all the production characteristics to which genetic engineering might be applied, it appears that the most obvious one – that of improving the yield of food from a crop or animal – is actually one of the most difficult. One of the features of current genetic engineering techniques is that they work best with effects which are controlled by single genes. Most of the genetic characteristics associated with yield, for example by enhanced growth, are under the control of a very large number of genes, each of which has only a very small effect on overall yield. A few genes are known which have large effects on production traits, but in animals these would need to have an advantage of at least 5 to 10 per cent in economic merit in order to be competitive with traditional breeding techniques.[9]

Early attempts at manipulating growth in animals failed due to severe welfare problems, especially with the now notorious Beltsville pig,[10] altered with human growth hormone genes. These pigs suffered deleterious consequences from the presence of this gene including gastric ulcers, arthritis, dermatitis, and renal disease. This might suggest that one of the theoretical advantages of genetic engineering – the greater selectivity in enhancing one function – may only go so far without doing harm to the overall balance of the animal's metabolism. As a result this area has not been as amenable to genetic engineering as had been hoped.

Animal production in general will probably continue to rely on conventional selective breeding, but supplemented by an *indirect* application of genetic technology, marker-assisted selection. As the genome of an animal becomes better understood, particular genes can be identified which can act as markers for various traits the breeder wishes to improve but which are not amenable to direct genetic modification.[11]

One notable exception to this pattern is in fish breeding. Trials are being carried out on salmon which have been genetically engineered to achieve a much faster growth rate.[12] Reports also suggest that transgenic sheep with improved wool growth have been produced in New Zealand.[13] Another indirect use of genetic engineering is BST (Case Study 7), in which the milk yields in cows have been boosted by repeated injections of a bovine hormone which was produced by genetically engineered bacteria. The welfare implications of this are discussed in Chapter 4.

Reduced Vulnerability of Crops to Environmental Stresses

The environment in which crops are grown has a major effect on their productivity. During their life cycle crops are subject to a range of stresses which can reduce their potential yield. These include extremes of heat and cold, drought, mineral deficiency and toxic chemicals in the soil. Case Study 2 discusses the possibility of altering crops to be resistant to these abiotic stresses. These crops would be of special importance in relation to developing countries. Although research discoveries point to some possible applications against drought, salinity and frost, on the whole this potential has so far proved quite difficult to realise.

Increased Nutritional Qualities of Food Crops

The seeds of legumes and cereal grains provide around 70 per cent of human dietary protein requirement. Unfortunately, proteins from these sources do not provide a balanced diet because they are relatively deficient in certain amino acids required by humans. Genetic engineering could allow the modification of proteins in legumes and cereals to a form approximating more closely to the necessary balance of amino acids. This would reduce the absolute quantity of food needed and aid world food supply. Similar considerations apply to the diets of farm animals. Here, modification of the amino acid content of foods could reduce the amount of waste produced by animal husbandry systems. Given that in some countries animal wastes have become serious pollution problems, this would have certain environmental advantages. These potential developments are still at an early stage.

Increased Nutritional Qualities of Food from Animals

Milk production is an area where genetic techniques have been specially applied. In The Netherlands, genetic modification has produced a line of transgenic cattle in which the milk composition has been altered to contain the human protein lactoferrin, which is believed to make it more digestible for babies and patients on antibiotics.[14] In the UK and US, the techniques developed for producing pharmaceutical proteins in sheep's milk are also being adapted for similar enhancements of the nutritional qualities of milk in cattle.[15] The welfare and other implications of these procedures are discussed in Chapters 4 and 5.

Improved Taste, Texture or Appearance of Food

In addition to the nutritional value of food, its taste, texture and appearance all affect its usefulness and attractiveness to customers. Many of the foods which are sold in shops and supermarkets suffer losses and degradation of quality during transport and distribution. This has always been something of a problem but is especially a feature of more centralised systems in which the areas where food is produced or processed are a long way from where it is sold and consumed. This is especially the case for more exotic foods such as tropical fruits which are transported over large distances. With fruit this often means that it is harvested before it is ripe, to reduce the chance that it will spoil before it reaches the consumer. Genetic engineering can be used to slow down this process of spoilage, so that fruit can ripen longer on the plant and then be transported to the consumer and still have a reasonable shelf-life. The Flavr Savr™ tomato in Case Study 6 has become the prime example in which a gene has been disabled, reducing greatly the rate at which the tomato softens as it ripens, so that it can be left on the vine and picked when it is more ripe and thus tastier.

Environmental Applications

Genetic engineering presents a variety of ways in which it could be possible to reduce the use of chemicals in agriculture, which could dramatically reduce the environmental impacts of current intensive farming practices. It may enable us to use a more integrated approach to crop husbandry with less reliance on chemical intervention, or use genes which confer natural resistance to pests, weeds, or viral and fungal infections. By engineering herbicide resistance into crops, there could be a faster shift from broad-spectrum chemicals to more benign ones. Genetic modification could also be used to develop crops which can produce substitutes for fuel oils, mineral oils, detergents and feedstocks for plastics. Genetically engineered micro-organisms have great potential in cleaning up various forms of pollution.

Enhanced Resistance to Weeds, Pests and Diseases, Using Less Chemicals

This is the area which has received perhaps the most attention in plant genetic engineering. Food crops suffer from a wide variety of challenges from weeds, insect or nematode worm pests, and fungal, viral, bacterial and other diseases, which cause serious problems for the farmer. The usual method to deal with these has been, for approximately 50 years, to use various chemicals which kill the organisms responsible. The resulting dependence on herbicides, insecticides and fungicides has led to environmental and health concerns, and has become a contentious issue between the agrichemical industry and those favouring more traditional organic forms of agriculture. Genetic engineering methods offer a number of alternative ways to tackle the same problems, ostensibly in a more targeted fashion, which, according to its advocates, could reduce substantially the amount of aggressive biocide chemicals being used. Evidence is beginning to emerge which would seem to bear out this claim.[16]

Resistance to several different herbicides has been engineered into a variety of plants. The idea is that the field can then be sprayed with herbicide which will kill the weeds but not the crop plants. The effect is produced either by increasing the tolerance of the plant to the herbicide (as has been done with the herbicide Glyphosate), or by having the plant break down the herbicide (done for sulphonyl-urea compounds). A number of crops are naturally genetically resistant to specific herbicides, like atrazine. If the genes responsible were to be isolated from resistant species and transferred into susceptible crop species, using genetic engineering, then the resistance mechanism would also be transferred. Attempts to do this by conventional plant breeding have been unsuccessful to date. The transfer of resistance should allow the more targeted use of agrichemicals and therefore reduce the amount of herbicide used.

Similar situations exist for a number of crops in relation to pest and disease resistance. Introducing genes from other plants, micro-organisms or even the pest itself can confer crop resistance and so reduce the use of pesticide sprays. For example, insect resistance has been achieved through the use of genes which produce a natural insecticide normally in a soil-dwelling bacterium called *Bacillus thuringiensis* (*Bt*). Crystals of the toxin have been used since the 1950s as an organically approved pesticide sprayed or dusted onto plants, but this type of technique is not without controversy as increased use is suspected to increase the development of B.t. toxin-resistant insects. Strategies for reducing the development of resistance have been suggested, for example including non-resistant plants in the mixture sown, and using crop rotation. However, this would depend on the integrated action of many individual farmers and therefore would be difficult to control.[17]

Work has also been carried out into obtaining resistance to insects through other means. Case Study 5 describes an example of using a genetically engineered virus to control insect pests on crop plants.

These uses of genetic engineering have been controversial. Opponents suggest they will in reality increase the use of herbicides and pesticides, and represent a cynical ploy on the part of agribusiness to increase sales at a cost to the environment. Proponents argue that plants have been engineered to be resistant to the more benign herbicides and the effect will be that of product substitution, so that the less harmful herbicides will be used rather than the more harmful ones. Nevertheless, the effect seems certain to tie genetically modified seed sales to the sales of a particular herbicide. Most of the funding of this area of genetic engineering originates with chemical companies, who are unlikely to devote considerable resources to the development of seeds which have no need of their products.

Reduced Dependence on Fertilisers and Other Agrichemicals

In addition to herbicides and pesticides, fertilisers are the major chemical additive used in agriculture, adding essential elements to improve the production and viability of crops. Nitrogen is one of the most important elements controlling crop productivity. Over 60 million metric tonnes of nitrogenous fertilisers are applied annually world wide. This is projected to increase to 160 million metric tonnes in the foreseeable future. However, a number of economically important crop plants, such as Soya beans, clover and other legumes, have evolved a symbiotic relationship with soil bacteria of the group *Rhizobium* which allows them to extract nitrogen directly from the air. This means that they do not require the application of nitrogenous fertilisers. Theoretically, this nitrogen fixing facility could be extended to plants currently without it, using genetic engineering techniques.

In practice, the introduction of the ability to fix nitrogen into plants is difficult and very complex, involving at least 15 different genes. The aim is to introduce either the ability to fix nitrogen directly or the ability to form a symbiotic relationship with *Rhizobium*, but either result is some way off yet. If this is successful, it could have a major impact upon world food supply, but it is also possible that it may not turn out to be quite the 'philosopher's stone' that is hoped.

Another application to chemical processes in agriculture involves genetically engineering microbial products to increase their efficiency at modifying the acidity and other properties of silage and hay. These microbes can dramatically reduce the losses in storage after harvesting which are caused by contaminating organisms.

Production of Novel Substances in Crop Plants

Increasingly, genetic engineering is being applied to both plants and animals for novel uses other than food – this is covered in several of the case studies. Oilseed is mainly used at present for margarine and various other food oils, but genetic modification (Case Study 3) could extend the range of such crops to include the production of fatty acids for detergents, substitute fuels and petrochemicals. Many of these applications are at an early stage, but they offer the prospect of significant environmental benefits – in cutting our dependence on non-renewable fossil fuels, and reducing net greenhouse gas emissions, since the carbon dioxide emitted by burning a biofuel is only that which the plants had originally absorbed from the atmosphere. Case Study 4 indicates the considerable potential use of crops for the production of vaccines and other pharmaceuticals by means of genetically modified plant viruses, with obvious medical benefits. If these various applications eventually become viable on a large scale, they will raise important questions of land use next century, because significant quantities of agricultural land would need to be set aside for energy crops, raw materials and medicines, instead of food. Although in the present UK context of over-production leading to surplus land this would appear to be a welcome development, that situation may not always apply and very different circumstances exist in many other countries of the world. Moreover, as discussed in Chapter 9, some of the genetic modification of oilseeds may have significant adverse economic effects on developing countries.

Use of Bacteria in Pollution Control

Bacteria can now be genetically engineered with designer enzymes and whole metabolic pathways capable of degrading a wide range of highly toxic man-made environmental pollutants (such as chlorinated hydrocarbons, pesticides, waste from munitions factories) to less dangerous or shorter-lived chemicals, or to mop up toxic heavy metals; processes known collectively as bioremediation.

Medical and Veterinary Applications using Animals

One of the developments in genetic engineering often regarded as most exciting is the range of novel medical benefits derived from animals. This has become the most important area of animal genetic engineering.

Pharmaceutical Proteins in Animals

Perhaps the most advanced application to date has been the production of therapeutic proteins in the milk of sheep, goats and cattle by introducing genes of human origin into the animal. The leading example is given in Case Study 8. There appear to be excellent prospects for a wide range of proteins to be produced by such methods, some of which are difficult to obtain by other means.

Xenotransplantation

A second area is the controversial modification of pigs or other mammals to enable their hearts, kidneys and other organs to be used for transplant into humans, seeking to overcome the rapid and catastrophic rejection of tissue from a foreign species that otherwise occurs. This is being developed in response to the growth in demand for transplant organs which in many countries has far outstripped the practicable supply. Case Study 9 describes the main UK work in pigs. There are uncertainties about ethical acceptance by the public, and the future of this work is also unclear in the face of concerns about the risk that it could lead to animal diseases being transferred to humans.

Animal Models for Human Diseases

The most widespread of all animal applications, however, has become the use of genetically modified mice as models for human disease and to test potential therapies. Partly because of the existence of embryo stem cell technology in mice, the application of genetic engineering to mice has been greater than to other animal species. Many different transgenic mice have been produced which have been genetically manipulated to produce various different diseases which are found in humans. The most famous of these mice is the oncomouse which has been genetically engineered to have a predisposition to develop cancer, described in Case Study 10.

Animal Disease Resistance

It may be possible to use genetic engineering to assist in combating some animal diseases, and to improve the resistance of farm animals to some diseases. This is a relatively undeveloped area, partly because the genetic traits associated with animal disease resistance, while of potentially great value, are often dependent on several genes. Possible applications include cattle with increased resistance to mastitis or sheep which have an insecticide against blow flies in their wool. A postulated application of animal cloning could be to produce a number of identical cloned animals which could then be compared in different environments in order to evaluate their susceptibility to diseases such as mastitis in cattle.[18]

Economic and Social Benefits

The governments of industrialised nations have long seen biotechnology as a rich potential source of wealth and job creation, akin to the information technology revolution. Biotechnology already employs many tens of thousands of people and may generate substantial profits for companies and their shareholders. This in turn should create wealth in the wider society when they are spent or reinvested. Much of this potential is still a long way off, however. High capital investment in research and extensive regulations for product testing and validation mean long lead times from the much heralded breakthrough in science to a company beginning to see profits from the product. It remains to be seen to what extent the economic benefits reach the wider society and if they are fairly distributed. There could also be economic benefits for more marginal agricultural areas of the world if genetic modification enables products to be grown on poor yielding land. On the other hand, if it becomes possible to grow, for example, palm oil substitutes on temperate farms in richer nations, hard-pressed tropical economies could suffer.

The field of genetics also presents many fascinating research problems for the scientific expertise and intellectual development of a country. The understanding which genetic research is bringing to the way living organisms work is a major addition to human knowledge, notwithstanding the ethical challenges this inevitably also brings, and which this book seeks to explore.

2 CASE STUDIES

Case Study 1

Lighting Up the Soil:
Genetically Modified Soil Bacteria

David Atkinson

INTRODUCTION

Pseudomonas fluorescens is a common soil bacterium which has been genetically modified to contain a gene which enables it to produce light under certain conditions.[19] The genes which cause this to happen are commonly found in marine bacteria and in a number of higher organisms of which the most widely known is the firefly. The development of this particular bacterium has been extremely useful to scientists studying basic soil processes, because the ability to produce light provides a tool which allows this specific micro-organism to be detected in the presence of other organisms. Furthermore, since light can only be emitted by active, living organisms, luminescence can be used as an indicator of activity in relatively natural environments. Much of the impetus for the development of such organisms has arisen because of the commercial potential to use genetically engineered micro-organisms in the environment, both for cleaning up pollutants and for controlling plant diseases. Consequently, there is a need to understand what happens to genetically engineered bacteria once they are released. This genetically engineered soil bacterium allows the investigation of microbial survival, growth, activity and dispersal within the environment, the persistence of the recombinant DNA and the potential for its transfer to micro-flora indigenous to the soil.

Main Ethical Issues Raised

Is There Something Inherently Wrong With Genetic Engineering?

This case study provides a situation where the question of the inherent rightness or wrongness of genetic engineering can be asked in its simplest form. Here is a case of genetically engineering an organism for whose fate there would be little or no public concern. Should even this be done? The inherent question about the correctness of transferring genetic information between species which would not and could not occur under natural conditions is an important one. These issues are further considered in Chapter 3. It could be argued that in the context of bacteria, gene transfer is a frequent occurrence, but is this different if the gene is from a higher organism? Equally, it could be argued that this process is unnatural because farmers do not manipulate bacteria. However, the reality is that farmers do manipulate soil bacteria by their agronomic practices, even if they are not aware that they are doing so.

Harm to the Bacteria

The introduction of the gene seems to confer no selective advantage to the organism. In fact, under conditions of environmental stress, the presence of the gene seems likely to be disadvantageous. This effect, which potentially could relate to the diversion of energy to light production, seems rather to relate to the position of insertion of the genes into the genome. The bacteria involved are of widespread occurrence and are abundant in most soils. The death of a small number of genetically modified bacteria is unlikely to have a major effect upon the preservation of biodiversity. The death of the genetically modified bacteria will result in their inability to pass on their particular genes to successor generations. The genetic modification is thus disadvantageous to the individual but probably has little or no effect upon the total population of these bacteria.

Human and Environmental Risk

This particular modified bacterium seems unlikely to pose any health risks. However, given the absence of a full understanding of the effects of the transformation, the creation of any modified organisms cannot be assessed as totally risk free. It would be possible for a modification to render the engineered organism more competitive than the non-modified type under some conditions, but less competitive in others. This might provide, through selection pressure, the conditions under which major variations in overall populations could occur. The genetic modification might easily transfer to other organisms, spreading the risk or problems to them. The change to the organism then becomes one of an unequal balance between risks and benefits. All of the benefits lie with the engineers of the organisms, the risks with the public at large and with the organism. In situations where the balance of risk and benefit are unequal, questions about the ethical basis of the genetic engineering activity must be asked.

Risks from bacteria are generally perceived as quite large due to the uncontrollable and ubiquitous nature of bacteria. There are a number of 'what if' scenarios which could be raised in this context. For example, what if these became 'superbacteria' and pushed out the useful ones? This is unlikely in this case because the modification puts the engineered bacteria at a disadvantage. Alternatively, it would be convenient for the scientists if the bacteria survived longer, so the next step in development might be in that direction (see the analogous situation in Case Study 5). One of the key questions is to what extent should all the various scenarios be explored? Some are clearly worth exploring, but are they all, however unlikely? Unintended consequences will always surround any scientific activity, but any risk assessment should always be comparative. Alternative risks also exist, for example not finding out what happens to genetically engineered bacteria. Aspects of risk are explored further in Chapter 7.

Case Study 2

To Boldly Grow where no Crop has Grown Before:
Genetically Modifying Plants for Harsh Environments

Michael Wilson

INTRODUCTION

Abiotic stresses are challenges to a living organism caused by physical and chemical components of the environment. Extremes of climate, water supply, mineral deficiency and toxicity have all played a major role in governing the global distribution and viability of crop plants. Not surprisingly, crop species are traditionally grown in areas most suited to them. For example, in Europe, oats and rye are more suited to the cooler, wetter north. Barley and bread wheat have become dominant in the middle temperate regions, and macaroni wheat has become established in the dry Mediterranean south. The main driving force for diversification has been natural selection. Human intervention has dramatically changed the situation by selective breeding, and more recently by enhanced breeding and genetic engineering. Crop plants can be adapted to new environments, with many potential benefits. At least in theory, such methods could be applied to improve the viability of agriculture in the developing world which is currently vulnerable to such stresses as drought and saline conditions.

Abiotic Stress Tolerance by Classical Plant Breeding

The cultivated wheats which we now use evolved by the natural hybridization of wild grass species around 8000 years BC in the Middle East. At first, humans developed these by crossing within a species to produce new varieties. More recently, plant breeders have also produced new hybrids between species. In the small grain cereals the best example is triticale, a wheat/rye hybrid which combines the hardiness of rye with the quality characteristics of wheat. This new crop species is now well developed and replaces wheat in poor quality, marginal lands throughout the world. Salt tolerant wheats have been produced by hybridising *Triticum durum* with salt tolerant lines of the wild species *Aegilops squarrosa*. Natural crossing – whether within or between species – produces thousands of different combinations of chromosomes in the progeny. Not all of these are beneficial, and although one desirable trait may be gained, undesirable features must be bred out by repeated back-crossing to the most agronomically desirable parent. To ensure genetic stability and crop consistency, this process can take more than ten years and is expensive.

Cytogeneticists work at the level of the plant chromosomes. They can mark, select and screen for the transfer of single chromosomes or small chromosome segments. For example, the short arm of the rye 1R chromosome carries a number of disease resistance genes and has been selected by plant breeders. It is now prevalent in wheat breeding programmes throughout the world. Cytogeneticists can also tag chromosome segments in species which carry genes for tolerance to a range of stresses, such as salt tolerance from *Thinopyrum bessarabicum*, waterlogging tolerance from *Agropyron repens*, and aluminium tolerance from *Secale cereale*.

The precision of this work is dependent on the detailed genetic maps which have been constructed for all the major crop species. The position of a gene can be located and genetic markers associated with it identified. These genetic markers are then exploited in breeding programmes to monitor and select for the incorporation of small segments of the donor genome carrying targeted genes. A disadvantage in incorporating whole or large segments of chromosomes is that undesirable linked genes can also be transferred. In genetic engineering only very small pieces of DNA, those carrying one or two desired genes, are introduced with minimal disruption of the rest of the recipient genome.

All the techniques described above exploit pre-existing variation within a species, or between closely related species. If the desired variation does not exist, it can be created through somaclonal variation (genetic instability which often occurs during plant tissue culture) or by exposure of seed to chemical or physical mutagens such as gamma rays. For example, the barley variety Golden Promise was produced by irradiation of the cultivar Maythorpe, a tall malting quality barley, to produce a short, high yielding malting quality barley. Golden Promise was the dominant barley variety in Northern Britain in the 1970s and 1980s, and is unprecedented as a cereal variety in being recommended for use for over 20 years. In addition to being shorter, the mutation process also reduced the sodium levels in the leaves of salt stressed barley plants. Similar radiation or chemical mutation methods have produced salt and cold tolerant rice, and aluminium and acid soil tolerant wheat.

Abiotic Stress Tolerance by Genetic Engineering

Acidity and Aluminium Tolerance

More than a third of the world's arable land suffers from soil acidity and aluminium toxicity. In acidic soils, dissolved aluminium injures the root cells in crop plants, interfering with root growth and nutrient uptake. It can greatly reduce the productivity of corn and other crops, especially in the tropics. Countering this by liming the soil is expensive for farmers and adds to pollution in run-off water. Mexican scientists have shown substantially increased root growth in tobacco and papaya, by introducing a citrate synthase gene from the bacterium *Pseudomonas aeruginosa*. As a result, the roots release much more citric acid which causes aluminium to be bound and immobilised as a citrate complex. If field trials with similarly modified rice and corn show comparable aluminum tolerance, this could have a major impact on raising agricultural production in developing countries.[20,21]

Frost Tolerance

Crop losses due to frost are estimated at over $3 billion per year in the US alone. To protect the crops, various physical methods commonly used include running wind machines, burning fossil fuels nearby and spraying crops with large amounts of water. These are generally costly and environmentally undesirable. One solution could be to introduce a gene from an Arctic fish which makes a special 'anti-freeze' protein which prevents the formation of damaging ice crystals inside cells. This could enhance frost tolerance in temperate zone crop species. An alternative approach modifies a bacterial gene which controls the production of a particular surface protein that acts to form ice crystals on crop leaves. Genetic engineering techniques have been used to produce bacteria in which this 'ice nucleation' gene has been deleted. By spraying these 'ice-minus' bacteria onto crops they compete with the unmodified ice nucleating bacteria and frost damage can be reduced. Curiously, the same machinery and ice nucleating bacteria are used in making artificial snow for winter sports resorts.

Drought and Salt Tolerance[22]

Recent predictions of the climatic effects of global warming anticipate an increase in extremes of weather, so that droughts and famines, already a serious problem for many areas of the globe, will become more widespread. Unless performed properly, with good topsoil drainage, irrigation is not the answer. It causes the soil to become salty, either by evaporating and leaving dissolved salts behind, or by drawing salt up from lower soil layers. Thus crop yields actually fall as irrigation and salinity increase. Between 1.5 and 3 million square miles of the Earth's land surface (about the area of the US) are defined as having saline soils and about 1 million square miles (ten times the area of the UK) are irrigated. About one-third of this land is becoming saline and so unable to produce crops at maximum yield or at all. Adequate drainage is too expensive to install for countries like India or Pakistan where the problem of saline soils is acute, and irrigation with salt-free pure water is simply not feasible or economical.

Sugar beet is the only crop plant which can tolerate modest amounts of soil salinity. Barley is less tolerant, despite many efforts to introduce this characteristic through sophisticated breeding and selection programmes. Until a few years ago, geneticists and plant physiologists dismissed the possibility of genetically engineering plants to tolerate high salinity or drought because these complex traits were known to be controlled by at least 100 genes in plants which naturally survive in salty or dry conditions. Genetic manipulation of so many disconnected genes at once is technically impossible at present.

Three years ago, however, two single dominant genes were identified and modified, one in each of two fungi, either of which conferred greater salt tolerance. Clearly these are candidates for future transfer to crop plants to see if they can improve resistance to soil salinity. In 1996, another potentially significant (if still somewhat scientifically controversial) discovery was made. A simple, single gene, which coded for an enzyme that synthesises the sugar trehalose, was genetically engineered into a tobacco plant, and the plant became drought resistant. Trehalose is a compound which naturally facilitates the uptake and retention of water and which stabilises other enzymes and proteins from the effects of dehydration.

General Tolerance

A more general stress resistance mechanism would be to use genetic manipulation to enhance the expression of natural, pre-existing plant genes known to be involved in stress tolerance. Enzymes such as super-oxide dismutase and peroxidase, which scavenge free radicals inside cells, are normally switched on by stress or damage, and act to protect plant cells by defusing these toxic chemicals. By changing their promoter sequences these genes can be made to express continuously and the plant defence mechanism is therefore in place before the stress is encountered. The transformed plants would then perform better under stress.

MAIN ETHICAL ISSUES RAISED

Underlying Issues

A whole range of techniques has been used to breed plants. Is there a point at which we feel we are going over the boundary of what is permissible to do? If so, where should this line be drawn? At selection within a species? At selection between crosses among related species? Identifying parts of the chromosome associated with particular traits? Identifying particular genetic markers for particular traits? Inducing genetic change by use of chemicals or irradiation? Using genetic engineering techniques to modify the regulation of a native plant gene? Using genetic engineering to introduce genes from another plant species? Using genetic engineering to introduce a gene from a bacterium, a fungus or an animal? Is genetic engineering only marginally different from other techniques? Is it justifiable to use our intellectual talents in this way in order to feed the world? These issues are considered in more detail in Chapter 3.

Environmental Impact and Biodiversity

If successful, the above approaches will increase the land area available for monocultures of genetically homogeneous crops, like drought or saline resistant wheat. This would lead to the removal of natural ecosystems and a reduction in the intrinsic biodiversity. Increased tillage of poorer soils may also lead to more erosion. Is this environmental cost justified in order to feed the growing world population? Should we be concentrating new technologies on further raising output levels from existing arable land to reduce the pressure to cultivate and destroy wilderness or marginal land ecosystems? Issues related to developing countries are considered in Chapter 9 and environmental risks in Chapter 7.

Commercial Driving Factors

Transgenic crops for marginal land in poor countries have low commercial interest, which means little money may be invested in producing them by commercial companies. Agencies which have been set up for research in developing countries and public sector and World Bank research centres are the major sponsors working in this area.

Case Study 3

A Thousand and One Uses for Oilseed Rape:
Novel Oils from Genetically Modified Oilseed Rape

David Atkinson

INTRODUCTION

At the present time, the UK supports an area of oilseed rape of around 420,000 hectares, most of this production destined for the food market. Around 16,000 hectares is cropped with high erucic acid rapeseed, which is used in the manufacture of cling film and in the production of a slip agent which allows polythene to flow freely. Within the world economy there is a multi-million pound market for non-food fatty acids which are the main constituents of the oil. Currently these are provided by the petro-chemical industry, but there is an opportunity for them to be provided by European agriculture. In addition, the UK alone requires substantial quantities of fatty acids, such as lauric and stearic acid, which are currently derived from imported coconut and palm oil. The UK imports around 400,000 tons of palm oil and coconut oil, much of which is used for the production of detergents. Progress in producing varieties of oilseed rape which are particularly high in desirable fatty acids is largely limited by the range of extreme types normally found within the species. A number of other species of crop plants have useful oils but they cannot normally grow at northern latitudes. Changing the climatic range of these crops is likely to take a considerable period of time and to date progress in the introduction by conventional plant breeding methods of characteristics such as cold tolerance, to tropical plants, has not been promising.[23,24]

An outwardly attractive option is therefore to manipulate genetically a successful crop which is well adapted to northern latitudes, such as oilseed rape, so that it can produce novel types of oil. Currently, varieties of oilseed rape have been created by genetic engineering to have high contents of lauric or stearic acid. The oil profiles of these genetically modified oilseed rapes are therefore similar to those found in coconut and palm trees. In addition, research is currently resulting in oilseed rape varieties modified to contain a range of other fatty acids. The potential also exists to modify the oils of other crops, such as coriander. These developments are usually justified in relation to a reduction in the level of imports and the creation of an industry and jobs in the UK.

MAIN ETHICAL ISSUES RAISED

Underlying Issues

The genetic modifications employed to date merely result in a change in the fatty acid profile of the oils contained within the rapeseeds. The question can be asked as to why this plant species does not contain these oils in nature. Has there been evolutionary selection against these oil types in these species? Conversely, classical plant breeding has enhanced the production of lauric and stearic acid from oilseed rape, so are we doing anything new?

Environmental Risk

The introduction of novel genes to oilseed rape appears to have no adverse effects upon the performance of the recipient rape plants. It is not known what would be the effects upon other wild brassica species which may receive the genes inserted into oilseed rape through pollination. A further question is any effect on honey produced from this pollen. The identification of all possible risks is clearly impossible. On the other hand, an environmental benefit could be claimed in using oilseed rape rather than petrochemicals.

Human Risk

The cultivation of oilseed rape in the UK has risen from 200 hectares in 1972 to 400,000 hectares in 1992, largely due to encouragement through EU subsidies. Problems of allergic reactions by humans to oilseed rape are becoming apparent. How might these be affected if the oilseed rape is genetically engineered? Other crops could be considered as alternatives for supplying these oils directly or by genetic modification.

Developing Country Issues

The principal ethical issue concerns the potential for genetically modified crops to reduce the requirement of northern Europe to import the products of palm and coconut from developing countries. If the oilseed rape technology continues to develop, then it could enable countries in northern latitudes to produce their own entire requirement for oils. This could have a major role in the economy and welfare of the populations of developing countries which currently rely on producing these oils for hard currency. Alternatively, it could be argued that this might benefit their populations if the change allowed or encouraged them to use their crop production capacity to produce food and other materials for their own populations, rather than a cash crop for the international market.

The development of varieties with changed oil characteristics will tend to centre on commercial influence with the high technology, highly capitalised firms which have the resources to produce seed varieties. This use of technology may distort natural trade balances, especially as the industrial producers will be large, and probably international, and the growers small and locally based. The balance of risk and benefit is not equally distributed either at a UK or international level.

A change in the production quality characteristics of a temperate species through the use of genetic engineering techniques could, at least initially, result in a reduction in the quality of life and in some cases the total loss of livelihood of people living in disadvantaged areas of the world. This change is likely to be the direct consequence of the transfer of genes from one species to another species, a situation which is unlikely to occur through natural processes. This type of change is not new, however. When oilseed rape was first grown in the UK, it replaced imported soya bean oil to produce margarine and soap, thus changing trade patterns. The difference in this case is that, since there is no temperate species that can produce these oils without genetic engineering, it is the existence of the technology that results in this particular threat to developing countries. These issues are discussed further in Chapter 9.

Case Study 4

Vaccination Made Easy:
Proteins From Plants, Using Genetically Modified Plant Viruses

Michael Wilson

INTRODUCTION

Potentially far reaching medical and veterinary applications have been found for plant genetic engineering. One is based on genetically modifying plant viruses, so that when they act on the plant, the plant tissues produce high levels of novel proteins.[25,26,27] By appropriate choices, these can be diagnostic, prophylactic or therapeutic materials such as vaccines, hormones, antibodies, antibiotics, anti-cancer or anti-HIV drugs. It is also possible that such methods could be used to improve the nutritional value of some crops, if the virus was modified so that it encoded a protein that gave the plant a higher than normal content of rare or essential dietary amino acids.

Plant Viruses

Over 900 plant viruses have been described. Plant viruses are generally regarded as undesirable, but often unavoidable, parasites in nature. There are one or two exceptions, such as the highly prized leaf or petal colour variegation caused by viruses in ornamental plants, for example abutilon or tulip. Cumulatively, plant viruses afflict all agricultural and horticultural food crops, ornamental plant species, woody perennials and weedy land plants. Some viruses are very host-specific, while others have a very broad range of susceptible hosts, for example alfalfa mosaic virus infects more than 400 plant species. They are ubiquitous. Humans ingest and handle plant viruses continuously without ill effect. These viruses are pathogens of plants alone, and as such they elicit a wide variety of symptoms in their susceptible hosts. These can range from no visible effect through to severe damage, stunting or even plant death, yet they remain intrinsically benign entities to herbivorous animals.

Most plant viruses (93 per cent or more) contain RNA rather than DNA as their genetic material. None have been shown to interact with the heritable genetic material of the plant host, or to affect it directly. Thus their environmental effects are relatively transient. Once a plant has become infected, viruses are not easily or economically curable. Their incidence is controlled, however, by strict national plant health regulations (for example EU Plant Passports), careful propagation procedures and, where possible, by breeding crop varieties with some form of natural, heritable resistance.

The Technique

Some plant viruses multiply inside infected cells with very high efficiency. Up to 50 per cent of the total leaf protein can be virus coded proteins and over a million virus particles can accumulate inside a single plant cell. Because of this, they can be viewed as a simple, cheap and rapid route to produce 'foreign' proteins inside the plant. Instead of genetically engineering the plant DNA, a plant virus is genetically engineered and the plant infected with this new virus. The newly synthesized products can then be extracted from the extruded sap of the virus-infected plant. Alternatively the modification can be done in such a way that the product proteins remain attached to the daughter virus particles, which makes it easier to purify on an industrial scale.

An added advantage of preparing human or veterinary medical products from plants by this means is that there is a very low risk of co-purifying any animal or human pathogen, such as HIV or hepatitis virus. This is a problem when extracting medical proteins from blood products, or using animal or human cell lines in culture as sources. Any plant proteins inadvertently carried through the purification process are likely to be harmless, perhaps even nutritious!

Two strategies are currently being adopted to exploit plant viruses to produce new proteins, both of which can be seen to offer numerous advantages over more 'traditional' genetic engineering of plant chromosonal DNA. These include high yield, low cost, speed and low risk. Several commercial organisations are now using these approaches. Both approaches rely on having a full length DNA copy of the relevant plant virus, which is capable of being genetically engineered. The natural, full length virus sequence is cloned and genetically manipulated, usually in bacteria, to produce further DNA or (more commonly) RNA copies of itself. These are then used to infect the plant. Copying can be done in a test tube or inside the plant itself.

The first method aims to produce a foreign protein in a soluble form inside the plant cell. The gene for that protein is added into the viral genetic material at a site where it does not destroy the intrinsic infectivity and viability of the virus. The foreign gene usually also carries some viral regulatory sequences to promote expression of the protein during the normal infection cycle. In the second method, the gene for a foreign protein or a shorter peptide (in the case of vaccine production) can be fused to an existing virus gene in a way that does not destroy the function of the virus. This is commonly done by fusion to the virus coat protein which is produced in the largest quantities – 180–2000 copies are needed to reassemble each daughter virus particle. Thus the yield of the combined virus coat protein plus foreign protein can be very high.

In conventional plant DNA transformation, the level of expression of the foreign protein is typically less than 0.1 to 1 per cent of the total leaf protein. By using an efficient virus with a high copy number and an added foreign protein gene, the first method can produce levels up to 3 to 5 per cent of all leaf proteins, 7 to 21 days after plant inoculation. Suitable viruses include the tobacco mosaic virus, potato virus X, or cowpea mosaic virus. In the second method, however, when expressed as a fusion to the virus coat protein, the foreign protein can reach around 10 to 25 per cent of total leaf protein. If the protein has a very high commercial value, then the speed, flexibility, containment, economics and facility of this approach have obvious commercial attractions.

Although still somewhat in its infancy, this is a very flexible technology which offers the possibility of producing efficient, cheap, edible vaccines, industrial enzymes, diagnostic proteins, hormones, new classes of peptide antibiotics, blood clotting factors, and therapeutic proteins such as anti-HIV or anti-cancer agents.

Main Ethical Issues Raised

Environmental Risk

The plant viruses being used as gene vectors to date are not spread by insects, mites, nematodes or fungi, which are the common agents for spreading most viruses in nature, neither are they spread by seed or pollen. Since no interaction or gene exchange occurs with the DNA of the host plant, there is deemed to be no risk of any long-term effects on the crop germplasm or on biodiversity. Because of the high value attached to the products from these genetically engineered plant viruses, the crop to be cultivated and infected will usually be small, well supervised and monitored for containment. In the US, open field trials of several dozen hectares of tobacco infected with just such recombinant viruses have been monitored for over five years in North Carolina and Kentucky. No escape has been detected either of virus or foreign gene into the neighbouring fields, nor to susceptible 'bait' plants deliberately sown around the test plots. Nevertheless, releasing a genetically modified virus into the environment will require careful control.

Animal Welfare

One benefit would be the availability of cheaper veterinary products and vaccines, to immunize farm animals against major pathogens, such as rabies, foot-and-mouth disease virus, *Pasteurella* bacteria and respiratory and enteric viruses. These vaccines could even be in the form of edible plants. Opportunities also exist to use a foreign, virus-encoded protein to divert plant metabolism away from the production of normally toxic compounds (such as alkaloids and lectins) which cause nutritional disorders in grazing animals. Alternatively, the nutritional value of a marginal crop could be raised by having the virus make, for example, a methionine- or lysine-rich protein.

Developing Country Issues

Opportunities exist for very cheap and easily applied treatments against intestinal or other parasites or vaccines against malaria, HIV, measles, polio, etc, by using a virus of an edible local crop (such as banana or plantain) to carry immunity-inducing peptides. Alternatively, opportunities exist for antibiotic peptides, therapeutic proteins and a wide variety of other products. It should not require building a very high technology local extraction and purification system, or chilled storage and distribution.

Food Related Issues

Humans handle and eat virus-infected vegetables on a regular basis. There appears to be no intrinsic nutritional risk from the viral expression system.

Commercial Driving Factors

There are potentially very high profit margins to be made because of the low costs of plant growth facilities compared to animal or microbiological cell cultures. Normal non-sterile agricultural and horticultural plants can be used and they can be grown in open fields.

Land Use

Use of agricultural land for the production of these crops may further increase the pressure on land use. However, the areas used are likely to be small and may involve marginal or set-aside land.

Will it Deliver?

This technology has a lot of promise, but will there be unforeseen problems ahead? One particular question would be whether there are limits to the products that can be produced by plants, via the viruses. Will minor changes in structure of the proteins result from differences in biochemistry between plants and animals, possibly in the addition of carbohydrates to proteins, and will these differences matter?

The Ethics of Abandoning This Technology

This case study illustrates a situation which, at least at present, appears to offer very large benefits from its adoption. Would it be ethically right to abandon all genetic engineering and lose this potential?

Case Study 5

The Sting in the Cabbage:
Genetically Modified Insect Viruses as Pesticides

Joyce Tait

INTRODUCTION

Viral Pesticides Against Caterpillars

The Institute of Virology and Environmental Microbiology (IVEM) in Oxford
has carried out an extended research programme with a view to developing
a genetically modified pesticide active against caterpillars. These caterpillars
cause considerable damage to highly valued commercial crops. At present,
frequent applications of toxic chemical insecticides are used to prevent this
damage, with adverse impacts on the natural environment and in some
cases also on human health. A biological alternative, with a narrower
spectrum of action and hence less environmental impact along with a better
human safety record, is therefore an attractive proposition.

The Oxford group based their work on genetically engineering an
insect virus, called the nuclear polyhedrosis virus (or NP virus for short)
which could then be sprayed onto a crop. The NP virus begins to work
when a sufficient quantity of virus particles are ingested by a caterpillar.
The viral coat is digested in the insect's gut and the virus particles multiply, causing eventual liquefaction of the caterpillar. The process of
liquefaction releases very large numbers of new viral particles into the
environment, ready to proliferate the infection in the caterpillar population. Each such cycle takes several days, during which time the caterpillars
can cause considerable damage to a crop, counteracting, for the farmer,
the attractions outlined in the previous paragraph.

More and Less Virulent Variations

As part of the risk assessment process for NP virus, the early genetic
modifications were designed to demonstrate that scientists were able to
track genetically modified viruses released into the environment, through
the addition of a harmless, readily identifiable genetic marker which

would enable the GMO to be distinguished from the unmodified (wild type) virus. As a further safety measure, the virus was subsequently modified to restrict its survival in the environment by deleting one of the genes responsible for production of the coat protein. However, in field trials, this weakened version of the virus proved much less effective as a pesticide than the wild type. This area of research was then set aside in favour of work designed to improve the pesticidal effectiveness of the virus and to extend its host range (both of which were seen as prerequisites for commercial success).

In subsequent experimental releases, the coat protein gene was retained to restore the virus to its original virulence. In addition, the NP virus was genetically modified to express an insect-selective toxin gene derived from scorpions. This considerably reduced the time taken to kill the host insect, and was claimed not to affect the pathogenicity or host range compared to the wild type virus.

Risk Assessment[28]

When IVEM applied for permission to carry out a trial release to test the pesticidal effectiveness of the virus incorporating the scorpion toxin gene, environmental groups tried unsuccessfully to prevent it. They raised questions about the possible impact of this virus on biodiversity. For example the virus could spread beyond the release sites, and persist in the environment from one year to the next. It could manifest virulence against non-pest caterpillars or it could exchange genetic material with other viruses capable of attacking non-pest species. The fact that the trial release of the modified organism in 1994 took place close to Wytham Wood nature reserve, close to Oxford, which is noted for some rare species of moth, exacerbated the conflict surrounding these questions.

The scientific aspects of these questions are still being debated. For example, based on the data collected during a field trial, although the modified NP virus killed the target species more quickly, the dead caterpillars did not liquefy and the subsequent rate of infection was lower than for the wild type. This has been interpreted to imply that the modified virus is less likely than the wild type to persist in the environment.

The early stages of risk assessment for this product investigated the spectrum of action and persistence of NP viruses that were specifically modified so as to present a minimal environmental threat should they escape from the experimental containment. However, there has not been a smooth and logical progression from these earlier trial releases. They are not a valid basis for providing reassurance of the safety of the viruses subsequently modified, as a result of commercial pressures, to be more persistent, to act more rapidly and to be more toxic to a wider range of species.

If or when products of this nature are available for commercial use, they are likely to be used on several major crops, for a range of pests in a number of different climatic zones. The delicate balance between persistence and decay is unlikely to be the same for each set of circumstances and so the risk of a GMO of this type remaining viable for more than one growing season, and therefore having the potential to cause environmental problems somewhere in the world, increases considerably.

MAIN ETHICAL ISSUES RAISED

Underlying Issues

To transfer the ability to produce a toxin from an arthropod to a virus which would not naturally have such a capacity, might be thought to represent a violation of species barriers, in an absolute sense, or else an inappropriate transfer of function. Is it a valid human intervention to give one species a quite unrelated characteristic of another? This is amplified by the fact that a food crop is involved and by the association of danger which scorpions normally have for human beings.

Commercial Driving Factors

The original motivation of the scientists developing this product was to produce an alternative to the chemical pesticides whose broad and indiscriminate mode of action can cause environmental damage. On the face of it, these are laudable aims in terms of both ethics and ecology. A number of concerns are, however, raised. These are primarily in terms of risk engendered by the nature of the toxin, by the use of a virus, and by the genetic modification itself, but they also touch on the influence of commercial pressures on genetic developments.

This case challenges the received wisdom about genetically modified biopesticides – that they would have a narrower spectrum of action, and thus be safer to use, and could be a means to stem the tide of chemically dependent approaches to agriculture. In practice, the turn of events surrounding the use of the modified NP virus shows that commercial pressures for a more widely applicable pesticide can cut across the technical objective, to the point where the original ethical justification for the product becomes compromised or even removed. Moreover, with all pesticides a balance has to be struck between the persistence and decay of the active agent, which in turn demands a balance of commercial and environmental considerations. At what point would legitimate economic concerns be said to have gone too far?

Risk-Related Issues

In the introduction to this case study, a number of environmental risks were identified which in large measure set the context for this case. It has also illustrated how commercial pressures can transform a product designed to overcome the defects of earlier generations of technology into one that is at least as risky or perhaps more so. From an ethical point of view the primary question from this is how a society controls the development of a new technology in which substantial changes to some species may have unintended consequences for other species and the wider environment. The constructions of risk made by different actors in the technology can lead to profoundly different conclusions as to the correct way to proceed. These constructions are often governed by the stress laid on various factors in the case including economic viability, environmental integrity, food production, food safety, scientific evidence, uncertain effects, expert evaluations and public perceptions. What represents an adequate description and analysis of a risk?

This question is focused in the UK and EU on the debate between precautionary and reactive approaches to the regulation of risk in biotechnology. For all its advances, genetics is still a relatively young science, and much fundamental knowledge is lacking in related areas such as soil science. Given the many unknowns, therefore, is it more appropriate to assume there are greater hazards than have been so far identified and only to allow the minimum of field releases? On the other hand, can a point be identified at which enough understanding has been gained, so that this precautionary approach can be relaxed, and what are the criteria for making such a judgement? This in turn relates to matters of public trust and public alarm. The dilemma is where the dividing line should fall between these two extremes, in order to deliver potential benefits from new technology while avoiding the potential risks. Issues of risk are further considered in Chapter 7.

Case Study 6
Genetically Modified Tomatoes:
Seeking Firmer Tomatoes with Better Flavour

Michael Wilson

INTRODUCTION

The Flavr Savr™ tomato was a pioneering example of a genetically modified food. In the US, most supermarket tomatoes are picked from the vine while still green and firm. This allows time to ship, handle and display them for sale before they go soft and eventually rot. They are artificially ripened by spraying with a chemical which releases ethylene (which is a natural plant hormone), but this does not increase the flavour. They appear red and ripe, but are actually hard, and tend to lack flavour. By comparison, 'garden-grown' tomatoes ripen on the vine and have more flavour. Ripening is due to an enzyme called polygalacturonase (PG) which degrades cell wall pectin, and which is produced by the tomato itself.

Calgene, a California based biotechnology company, isolated the DNA which encodes tomato PG. They then genetically reversed a copy of it, making an 'antisense' gene, which they reintroduced into tomatoes by genetic manipulation. The method chosen also involved incorporating an antibiotic (kanamycin) resistance gene (*kan'*), to select transgenic tissue from non-transgenic (see Chapter 1). The antisense PG DNA produced a messenger RNA inside the nucleus of the cell which bound to and inactivated the message of the natural ('sense') tomato PG gene. The effect was to reduce greatly the amount of PG enzyme made in the tomato, and so slow down the rate at which the tomato went soft. As a result the tomatoes could be left longer to ripen on the vine, acquiring more flavour and still have time to be shipped to supermarkets. This was marketed under the name Flavr Savr™ tomato.[29]

Development of the Tomato

For several years, Calgene and rivals DNA Plant Technology (Vinesweet™ tomato) had been marketing non-genetically engineered, vine-ripened tomatoes which were selected for low PG activity through somaclonal variation, a natural genetic mutation which arises during plant tissue culture procedures. Eventually, however, Calgene withdrew this product because of poor profits. There were a number of problems including lack of genetic stability, product predictability and variable quality. The withdrawal also undermined the competitor Vinesweet™ strategy and product. The first Flavr Savr™ tomatoes were sold in shops in Chicago and California on 30 May 1994. In 1997, however, Monsanto, who had now taken over Calgene, withdrew the Flavr Savr™ tomato from US markets. Ironically this was because the original variety chosen for the genetic modification was not a good variety for taste, texture and disease resistance. This illustrates that a genetically modified end-product still remains dependent on the quality of the original plant variety chosen for the modification. Although not a commercial success, the Flavr Savr™ tomato and its predecessor could be said to have prepared consumers for the idea of genetically modified food.

In parallel with this, the UK company, Zeneca Plant Science, developed a tomato paste made from transgenic tomatoes, based on suppression of the normal ('sense') PG gene, rather than using an antisense gene. This paste became available in some UK supermarkets (Sainsbury's and Safeway) in 1995 and has remained on the market. DNA Plant Technology Corporation and Zeneca aim to use the former's US-patented Transwitch™ sense gene suppression (not antisense) and Zeneca's low ethylene technology to develop bananas with delayed ripening properties. This would enable the transport of specialty bananas hitherto not suitable for prolonged shipping. Zeneca also aims to launch similar tomatoes, as well as slower ripening strawberries, melons and peaches.

Regulatory Approval

In 1989, although there was no mandatory US Federal approval requirement or regulatory process for genetically engineered foods, Calgene began voluntary consultations with the US Food and Drug Administration (FDA) to assess the safety of their tomato. Calgene believed that consumer opinion and acceptance would only come with full, transparent, informed and scientifically objective scrutiny – to give customers confidence and the information on which to make their own choice. The case became a landmark not only for being the first genetically engineered whole food to receive FDA scrutiny and reach the market, but also for setting the guidelines for further future (voluntary) FDA applications for transgenic foods. The FDA concluded that Flavr Savr™ tomatoes did not differ significantly from traditionally bred tomato varieties.[30] They were functionally unchanged except for the intended effects of the antisense PG gene and the antibiotic resistant *kan^r* gene. As such, they remained a food and subject to regulation as a food. Even critics have urged the FDA not to impose such prolonged, strict and cumbersome procedures as were directed at Calgene and the Flavr Savr™ tomato during its five-year voluntary review. Subsequent cases have generally moved faster and involved less rigorous investigation, at least in the US.

There was extensive patent litigation over the antisense PG tomato gene, involving Calgene, Zeneca and Enzo. The Campbell Soup Company funded both Calgene and DNA Plant Technology, respectively, for Flavr Savr™ and Vinesweet™, and hold the rights to both products, and to process the Flavr Savr™ tomato. Calgene was concerned about patent infringements by amateur gardeners growing its modified tomatoes from seed at home and about transgene escapes (in the UK product, processing would destroy tomato seed viability). See Chapter 8 for further details on patenting.

MAIN ETHICAL ISSUES

Underlying Issues

The tomato technology poses a basic question whether genetic modification represents any greater manipulation of nature than chemical alterations. In so far as the antisense tomato PG gene was engineered from a natural piece of sense tomato DNA, and was added *via* an artificially disarmed *Agrobacterium tumefaciens* plasmid (see Chapter 1), a natural biological event in the lifestyle of these soil bacteria, there was significant human contrivance in the final living product. This was the basis for it being patentable in the US. It could be argued that a more natural alternative method to select similar low PG plants is by random somaclonal variation, which does not introduce such a highly targeted, selective and specific DNA sequence. This approach, however, tends to affect more than a single trait. It can be genetically unstable and have multiple side effects. Although arguably more 'natural', it still requires the unnatural production of daughter plants, through single cell vegetative propagation in artificial culture media with a significant, skilled intervention by humans. It is questionable that this is more manipulative of nature than existing methods like using an ethylene spray to ripen tomatoes. People would not eat green tomatoes. Is it more ethically acceptable to eat prematurely picked tomatoes that appear ripe and red through an artificial chemical manipulation than to eat tomatoes whose natural ripening process is delayed genetically?

Risk to Humans

During the FDA's assessment, the safety concerns shifted from the PG antisense gene (which is chemically and biologically just another piece of tomato DNA) towards the product of the kan^r selectable marker gene (0.1 per cent of total tomato protein) which was used during addition of the foreign DNA. Calgene made a formal Food Additive Petition to the FDA in 1991 to deal with speculative risks that this antibiotic resistance gene could compromise public health. After extensive tomato feeding experiments with rats, the FDA and its Food Advisory Committee declared the kan^r gene safe. The protein which confers kanamycin resistance is not a toxin or allergen, and its gene is highly unlikely to be transferred horizontally from plant DNA to soil or gut micro-organisms. It would not compromise antibiotic use in humans by transfer to gut bacteria. A wide range of trials were made including molecular analyses, biochemical analyses, nutritional studies (eg levels of vitamins A and C), horticultural traits, genetic analyses, field trial results, plant pest risk evaluations (foreign DNA or potential 'weediness'), endogenous levels of potentially toxic compounds (tomatine and solanine), and taste. All the data from these indicated that Flavr Savr™ tomatoes were equivalent to normal tomatoes and posed no human health or environmental safety risk.[31]

Such data, ironically, raise an important question in precautionary risk assessment. The need to prove a negative, ie that there is no risk, is impossible to meet. While the technical risk may be very low, the perceived risk by the public may be much higher if it involves something to which people tend to be averse. The result can be a gulf between scientists and public, analysis and perception. This is discussed further in Chapter 7. In the light of this, plant breeders are now tending to remove antibiotic resistance genes prior to cultivation on a large scale.

Environmental Impact

In 1992 the US Department of Agriculture also concluded that the Flavr Savr™ tomato posed no risk to the environment. There is still concern, for example from the US Union of Concerned Scientists, about the widespread use of antibiotic resistance gene markers for possible events in plant debris which are broken down by bacteria in soils. To date, all risk experiments have come up negative.

Developing Country Issues

The same technology may be used to produce exotic fruits and vegetables with delayed ripening. This may encourage their growth in developing countries and their export to industrialised countries. This could provide further opportunities for cash crops and wealth creation, or else further problems from undesirable effects involved, as considered in Chapter 9.

Food Related Issues

The FDA consensus view was that 'use of an antisense gene to block PG production posed no conceivable risk to consumers'. Although impressed by the safety data on the Flavr Savr™, the lack of obligation to provide such data led to major criticisms of the adequacy of the FDA's oversight procedures for biotech foods in general, and of the product per se, as for milk produced with BST. Jeremy Rifkin of the Pure Foods Campaign called for a 'tomato war'. Calgene voluntarily labelled the tomatoes as being genetically engineered, although process labelling was not required. It had plans to explain the technology in brochures available in shops.

A different issue arises from processed foods using these modified tomatoes. The Campbell Soup Company and Zeneca use genetically engineered tomatoes in various products. Should all these derivative products be labelled also? Zeneca labelled their tomato paste derived from transgenic tomatoes, but the wider picture changed once US genetically engineered soya came on to the European market unsegregated from mixtures of unmodified varieties, and unlabelled. This and other food issues are assessed in Chapter 6.

Commercial Drivers

The market for this type of tomato is created by the way in which society arranges its food supply. Consumer demand for variety, quality and year-round availability of seasonal produce is important to major retailers. Longer shelf life is an attraction when the product has to be transported long distances all year round from where it is grown to where it is eaten. This is especially important in the US. In the UK this system is being questioned on environmental grounds, with trends towards using more local produce being encouraged.

Calgene and others perceived the potential of antisense technology in the mid-1980s, when it was shown to be a natural means for control of gene expression in bacteria. PG was a suitable target in a high value cash crop with marketing problems looking for a solution. Yet Calgene existed for more than ten years and never showed a profit. With the US fresh tomato market estimated at $3 to 5 billion a year, Calgene's Flavr Savr™ tomato might have generated huge revenues. The fact that it did not was partly due to the lengthy regulatory process and its associated costs and partly because the particular tomato variety which was chosen initially (brand name MacGregor's) was withdrawn because it was agronomically poor.

Case Study 7

Bovine Somatotrophin (BST):
Boosting Milk Yields with Hormones Produced in Genetically Modified Bacteria

Peter Wilson

INTRODUCTION

Bovine somatotrophin (BST) is a naturally occurring hormone of the cow, secreted by the pituitary gland. As the name implies, it regulates growth of the animal including the growth of the secretory tissue of the mammary gland; hence the administering of a larger amount of BST than normal, in the mature cow, leads to the output of significantly greater quantities of milk.

BST can be synthesised by using recombinant DNA technology. In practice this means using the organism *E. coli* (a common bacterium of the gut) and altering its physiology by genetic manipulation so that it makes significant quantities of BST in its cells. These cells can then be harvested and the BST extracted. The resultant product is chemically identical to natural BST, but as it is produced artificially by recombinant DNA technology it should strictly be known as rBST (the 'r' standing for 'recombinant DNA'). BST is not licensed for use in the EU but is used, albeit controversially, in many parts of the US and it is licensed, or used without a licence, in a number of other countries.

MAIN ETHICAL ISSUES RAISED

Effect on Cows of Higher Milk Yields

BST is certainly efficacious. It raises milk yields by a very marked 6 to 25 per cent, and it does so at a greater efficiency of feed conversion. Thus, although BST-treated cows eat more, they eat proportionately less extra food than that required for milk production in non-treated cows. BST's efficacy is estimated to be worth about eight generations of selective breeding (about 24 years).

This raises the question as to whether these higher milk yields are harmful to the cow. With very well managed cows there does not appear to be any ill effect (for example shorter life; increased mastitis; more lameness etc) but in poorly or badly managed cows the extra milk output adds to the stress on the system and in such cases more ill effects have been noted. So the administering of BST presupposes that the cows receiving the treatment are managed in an above average way. This will not always be the case.

Land Use

It follows from the above that it is possible for a farmer (or a nation) to produce the same amount of milk from fewer cows (since yields are significantly increased). This has environmentally advantageous effects in that cow numbers could be reduced, with less pressure on the land and thus less pollution. Such considerations could be very important in over-populated countries such as Holland and many overcrowded developing countries such as India and some West Indian islands.

Non-Therapeutic Injections

BST can only at present be administered to cows by injection, and repeated injections are currently needed. In the early experiments daily injections were used, but the pharmaceutical companies are working on long-lasting forms of preparation so that less frequent injections may be possible in future, such as weekly or monthly. However, the injection method does raise ethical questions. Most people (and all vets) are happy to use injections as part of therapy, to cure illness or prevent disease. Such people would be less happy to justify injections purely in order to give greater economy in milk production. There will be many vets who would refuse to carry out such injections if asked – farmers may be less reluctant to do so if the economic benefits are tangible.

Exploitation of the Cow

Some would hold that there is an 'in principle' objection to increasing the pressure on a cow to increase the amount of milk she produces. However, selective breeding of cows has increased, and will certainly continue to increase, the mean milk yield of cows, goats and buffaloes, hence a refusal to use BST will not do much, if anything, to reduce the pressure on the cow to produce more milk.

Safety of the Milk

Two aspects of safety have been raised. One relates to the potential for BST administration to result in higher incidence of mastitis. This in turn will require treatment with antibiotics at the present time and therefore opens up the possibility of increased antibiotic residues in milk.

The second issue relates to the fact that BST injection results in increases in the amount of insulin-like growth factor-1 in the milk of treated cows. It has been suggested that various health problems might increase due to this. However as the claimed effects are small, it is difficult to prove or disprove this hypothesis.

Licensing Criteria

BST is not presently licensed for use in the UK or the EU and the failure to award a licence is based more on political and marketing grounds than scientific evidence. Opponents argue, for example, that there is no need for the product, given the surplus of milk in Europe, and that it will harm smaller farmers. If only the scientific evidence was cited the Veterinary Products Committee would by now have awarded a licence as the scientific case is robust; in other words, BST satisfies the three legal criteria for safety, efficacy and quality. This raises an interesting legal-ethical question; should the licensing authorities be swayed by public opinion when they are legally only required, according to the Medicines Act (1968), to consider the scientific evidence on safety, efficacy and quality (the only three criteria allowed to be considered under the Act)? Many people doubt the scientific reviews and accuse bodies who have carried out the research of being biased in favour of the manufacturers. Conspiracy theories have been suggested and claims have been made that the manufacturers have suppressed evidence which is not entirely favourable to their product. If the scientific and regulatory bodies are to be trusted, however, there appear to be no problems with the safety of BST.

Food Issues and Labelling

In the past, milk boards were obliged by law to buy all milk from producers which came up to the legal (safety) standards. As milk produced in part by BST administration did come up to these standards, such milk had therefore by law to be collected by the milk boards, pooled, processed and distributed. The consumer therefore did not know whether the milk or milk products purchased from the dairy had been, in part, produced by BST or not. Should the consumer have had such knowledge? Even if the answer is 'yes' it would have been very difficult, if not impossible, to collect and bottle BST milk separately from 'normal' milk. How should this logistic difficulty have been tackled? If the milk was 'safe' does it matter how or where it was produced? It is argued in Chapter 6 that food is of such fundamental importance to humans that labelling is important to allow people to make choices. Experts are not always as well informed as they think, so it seems reasonable to give people the choice, even if the risks are deemed to be low, in order to recognise people's right to exercise individual choices.

When the initial trials of BST were carried out in the UK it was done, quite legally, under cover of an Animal Test Certificate. This certificate allowed cows to be injected, and milk collected and distributed, before the trial results were known (in terms of efficacy and quality and safety to the animals). Should milk (or any other product) enter the food chain before all tests have been completed and licences granted? It might have been wiser to have required all such 'experimental' milk to be disposed of outside the normal milk market, such as by feeding back to pigs or calves.

Although rBST is chemically identical to 'normal' BST it is clearly a different product. It is produced by a different industrial process and it undergoes 'chemical separation and purification' which ordinary BST does not. Is it, therefore, morally and ethically correct to argue that rBST milk is equivalent to BST milk? The chemist (at present) may not be able to detect the difference but the perception of the two products is clearly different and there could conceivably be other physico-chemical differences as yet undetected.

Economics and Regulation

Whether or not the farmer uses BST if it is licenced will depend upon its economics. BST is not expensive to produce (it is a mass production factory type process) but the companies involved have spent many millions of pounds on research and development and will price it at a level that enables them to recover their initial costs and make a profit thereafter. If the price is too high farmers will not buy. If the price is pitched low it is likely that the product will be used either openly and legally or, in some countries, even illegally on the black market, which is already thriving on the sale of unlicensed and hence illegal drugs, especially in the less regulated countries. It must, however, be in the interests of the companies who have invested so much money in the research programme, to have their product universally trusted and accepted.

This case study of BST illustrates the way in which new products of genetic engineering (in this case a minor genetic alteration of *E. coli* bacteria) raise many far-reaching questions which are not at present being tackled as they are outwith the scope of existing laws and regulations. It is of interest that European ministers are at present opposing the licensing of rBST for 'political' reasons, possibly more in response to public opinion than in conformity with the powers vested in them by law. Are they correct or incorrect to operate in this way, which has no present legal sanction? In other words, should there always be room for an override of scientific evidence 'for the public good'? If this is desirable, the legislation should be altered to enable non-scientific considerations to be taken into account.

Case Study 8

Pharmaceuticals from Milk:
Producing Pharmaceuticals in Sheep Milk

Ian Wilmut

INTRODUCTION

One of the most important areas of genetic engineering in animals is the novel application in which proteins which could be of value in treating human diseases can be produced in the milk of farm animals. This was based on the observation that the regulatory elements of one gene may be used to direct production of a protein by an organ which does not normally have that capacity. Suitable regulatory elements of the gene must first be identified and then the correct genetic change has to be introduced into the animal. Case Study 11 refers to a new way in which this transfer is being attempted using nuclear transfer techniques. Once modified, success also depends upon the ability of the organ, in this case the mammary gland, to synthesise and secrete the relevant proteins in sufficient quantity.

A number of milk protein gene control sequences have been identified which direct expression of genes to the mammary gland. The concentration of protein in the milk varies considerably between individual animals, but several animals have now been bred which produce commercially useful concentrations. Most applications to date have been in sheep, goats and more recently also cattle. This case study focuses on the first product which has come to the stage of clinical trials, that of alpha–1–antitrypsin (AAT) in sheep's milk, pioneered by PPL Therapeutics plc and the Roslin Institute.[32] A number of other products are being produced in sheep, cattle and goat milk at various research organisations around the world.

The proteins of interest in this work are biologically active complex molecules. They carry sugar residue molecules which are added to the chain of amino acids making up the protein. These molecules are believed to be important in determining the length of time that the protein persists in the animal and also the manner of its interaction with other proteins and cells. This has presented a difficult problem in attempting to produce these by conventional pharmaceutical production. Standard methods using bacteria or yeast can only produce simple proteins which do not carry sugar residues, such as insulin. They do not appear to be able to synthesise the correct sugar groups of more complex proteins. At present such proteins have to be isolated from blood or human tissues. This is expensive and may carry the risk of infection. In some cases these are only available in limited amounts. In contrast the mammalian genetic engineering route offers a means to synthesise complex proteins in quite large quantities, with the correct sugars, and with less problem for purification.

AAT was the first protein to be obtained at high concentration in this way. It is needed to treat the lung disease emphysema and can also help in treating the symptoms of cystic fibrosis. It acts as an inhibitor of an enzyme which breaks down proteins in the lung. In emphysema a genetic defect causes insufficient AAT to be produced, and lung function can be seriously impaired. One of the first sheep to produce AAT in her milk in considerable quantities was known as Tracy. The second stage clinical trials with AAT began in 1997, with the aim of commercial production early in the next century. The present evidence is that all of the proteins secreted are biologically active and are very similar to the native human protein. However, in many cases there are small differences. Confirmation that the differences are not important enough to limit the use of the proteins is still awaited.

MAIN ETHICAL ISSUES RAISED

Mixing Genes Between Species

In this application, the fundamental objection to transferring genes between species receives one of its strongest challenges, given there is such obvious potential human benefit for a relatively minor intervention in the animals concerned.

Risks to Humans

Risks need to be considered both for the patient receiving treatment and for the general population. There are two risks for the patient. The proteins would typically be required at regular intervals over the lifetime of the patient. Although the very greatest care is taken over the health of the animals there is a possibility that an infectious agent may be introduced with the protein. Haemophiliacs receiving blood products have been infected in the past, for example. Secondly, if the proteins are slightly different from the native proteins there may be side effects. The biological effect may be slightly different and it must remain a remote possibility that the proteins will provoke an immune response.

The proteins are produced for medical use, but since they are in the milk of farm animals, food issues are also relevant. If either the milk or the carcass of a transgenic mammal were used as food, either inadvertently or after legal approval, would this present problems for either food safety or ethics? In the UK it was a related question which led to the setting up of the Polkinghorne Committee on the Ethics of Genetic Modification and Food Use. Issues concerning transgenic food are considered in Chapter 6.

Environmental Impact

The animals themselves are externally indistinguishable from others of the same breed, and so the relevant transgenic lines need to be kept separate from unmodified animals. Should some future need arise, there would therefore be the opportunity to identify and destroy the relevant transgenic population, but the possibility of accidental mixing can not be entirely ruled out.

Animal Welfare

All the indications to date are that the sheep with the AAT gene are healthy and normal. There are, however, two concerns. The present techniques involve surgery upon several donors and recipient animals to obtain each founder animal, perhaps 50 sheep. These animals suffer the anxiety of handling in unusual circumstances and pain subsequent to the operation. In almost all cases this will be by far the greatest suffering in such a scheme. Secondly, there is a possibility that protein may leak from the mammary gland and be active in the producing female. This can cause problems with some proteins. For example the protein erythropoeitin stimulates the production of red blood cells and the presence of an excess of this protein is expected to cause death. It would therefore not be a candidate for production in the mammary gland without using methods for the production of a biologically inactive form of the protein. Issues concerning animal welfare are considered further in Chapter 4.

Case Study 9
Xenotransplantation:
Organ Transplants from Genetically Modified Pigs

Ian Wilmut

INTRODUCTION

Xenotransplantation or xenografting is the transplantation of organs between different species. In view of the shortage of human transplant organs it has been proposed that certain organs like the heart, kidney and pancreas could be obtained for transplantation from pigs. In recent years the increase in organ transplantation has reached the point where demand now exceeds supply. This is despite efforts to encourage people to carry donor cards, and the adoption of management systems to ensure that organs recovered from human casualties are used more efficiently. Since patients who would benefit from organ transplantation are, by definition, critically ill, this inevitably means that some people will die before a suitable organ becomes available. It is in these circumstances that proposals have been made to use organs recovered from genetically modified pigs.[33,34,35]

Tissue Rejection

When an organ is transplanted, the body has two systems which may lead to rejection of the new tissue. If the transplantation is between two animals of the same species, the organ is rejected because of the presence of various molecules on the surface of the cells which the receiving body does not recognise. By far the most important of these are a very large family of molecules called histocompatibility antigens. Different members of this family are essential for the production of antibodies to destroy infective agents such as bacteria or viruses. This response is relatively slow and may require days or weeks to cause the death of the transplanted organ. It is countered by administering immunosuppressant drugs to the patient.

By contrast, xenotransplants are rejected by a hyperacute response mechanism which causes the death of the tissue within minutes or hours. The great speed of this response reflects the fact that there are preformed antibodies which react against tissues which are so very different. This is not because the antibodies are present to prevent 'invasion' by a transplanted organ, but apparently because sites on the tissue resemble those involved in the response to infection. The objective of the type of genetic manipulation considered in this case study is to prevent this hyperacute response. Other compatibility issues may also need to be addressed, possibly also calling for genetic modification steps. It is also likely that conventional drug therapy will still be required to prevent rejection by response to the histocompatibility antigens.

Cell death by hyperacute rejection is caused by a group of enzymes which bind themselves to the surface of the cell. These enzymes are present in blood serum and are known as 'complement', because they complement the activity of the antibody. They act together to create a large molecule which attacks the cell membrane, makes a hole through it and so leads to the death of the cell. There are protective mechanisms which prevent an animal's own complement from damaging its own cells, known as complement regulator mechanisms. These are, however, restricted to one species, so that human complement would not protect pig cells transferred into a human patient.

Genetic Manipulation

The aim of the genetic manipulation is to enable pigs to adopt the human complement regulator mechanism, so that if a pig organ is transplanted into a human patient, the body does not recognise it as being a foreign species. Genes which direct the production of the regulatory substances or factors have been transferred into pigs. The pigs then produce the human protein on their cells, in addition to their native pig protein. Research groups are working on several different regulatory factors. In Cambridge, pigs have been produced by standard gene transfer techniques (Chapter 1) which produce the protein as expected. Organs from these animals have been tested in tissue culture baths using human blood. Whereas a normal pig heart is very rapidly destroyed under such circumstances, the test hearts survive and function for longer periods, suggesting that the regulatory mechanisms are functioning as expected.

MAIN ETHICAL ISSUES RAISED

Xenotransplantation has proved a controversial idea, and two significant ethical studies have been undertaken in the UK. The first was by the Nuffield Council on Bioethics, published in 1996, and the second was commissioned by the Department of Health, published early in 1997. The recommendations of the latter report have resulted in a moratorium by the UK government on clinical trials for the time being, because of fears of disease transmission.

Underlying Issues

Is it Wrong in Principle to Use Animal Organs for Humans?

Animals have provided humanity with many things including food, clothing, traction, transport and companionship. Does the use of animals as donors of organs represent an unacceptable extension of these age-old uses? To some people it is wrong in principle. Against this, others question why it should be seen as any different to slaughter an animal to eat in order to live, than to slaughter it for spare parts to prolong human life. An interesting paradox is that we eat animals but not humans, but we are more dubious about having an animal organ transplant than a human one.

Is There a Problem with Mixing Human and Animal Tissue?

There is a crucial ethical point in the crossing of normal barriers between species. Is this something which we should not do? There are, for example, a number of Old Testament injunctions which prohibit mixing between 'kinds'; do these apply in this case? The specific use of pigs could also raise objections within Jewish and Islamic religious traditions. The concept of species is a human construct in an attempt to classify biological diversity of the natural world, but are there fundamental intuitions underlying this which should be respected? This issue is further explored in Chapter 3.

Is it Different From Using Pig Heart Valves?

Certain forms of heart surgery make use of valves taken from pig hearts. These have been specially treated to kill the tissue and prevent rejection. Is xenotransplantation just an extension of this, or does it differ ethically because the hearts would be complete living organs, and not just a particular moving part? Would it be different again from another development in xenotransplantation in which modified animal cells might be transferred rather than whole organs? Does size also make a difference? Some people may accept a small implant but would be more cautious about large quantities of tissue.

Risk to Humans

One of the dangers which is inherent in this type of research is that it is not possible to predict the outcome. Is it ethically right to do this? The first patients would face a very uncertain future. Since their alternative would presumably be imminent death, they might be content to take the risk, or see it as worthwhile in order to benefit those who come after them. The research so far has concentrated upon overcoming the immunological difficulties; however, problems may arise because of differences in the physiology of humans and pigs. But there are also differences in life-span, heart rate, blood pressure and the structure of regulatory hormones which maintain the basic physiological stability of an animal. It is not clear how an organ from one species would perform in another. Beyond the individual, the government moratorium has focused on the risk of transfer of disease from pigs into the human population as a whole. Though a remote prospect, current evidence indicates that it cannot be discounted completely.

Human Welfare Versus Animal Welfare

The desire to alleviate human suffering is a principle which all would wish to uphold, but the questions 'At what cost to the animal?' and 'To what benefit to the human?' must also be asked. The procedure is obviously highly invasive for the animal, in that it is being killed for its heart. Up to that point are there problematic welfare issues? The above need for a special disease free environment for the pigs could present especial problems for the quality of life of this inquisitive and intelligent animal. During the research phase, a number of experiments could raise welfare objections. Such issues are addressed in Chapter 5.

Case Study 10

Modelling Human Diseases:
Genetically Modified Mice as Models of Human Diseases

Donald Bruce

INTRODUCTION

By far the most extensive application of genetic engineering in animals to date is in the production of transgenic mice as human disease 'models'. Typically such mice have been genetically manipulated with a mutation which gives the mouse a propensity to develop a particular disease, for which the mouse thus becomes a model. In the case of 'knockout mice' a gene is disabled or removed, in order to discover its function. Mice are being used in this way to study a wide range of serious human diseases – the factors which trigger or arrest a disease, the way it develops, and so on – and to test the efficacy of potential therapeutic measures, including new chemical treatments and gene therapy.

Examples of Disease Model Mice

The use of animals as models has, for many years, been a central part of the development of understanding of disease and of testing new treatments. Over 2000 varieties of mice are currently available which have been selectively bred to enhance natural characteristics. It has been found that, for inherited disorders, in many cases very similar genes perform equivalent functions in mice and humans. Thus it is possible, virtually routinely, to introduce a specific genetic mutation into a mouse at the same gene that in humans leads to a particular inherited biochemical disorder, with the expectation that the mouse will develop the same disease. Mice have been produced which can mimic various forms of cancer, Duchenne muscular dystrophy, cystic fibrosis, Alzheimer's disease, and many other inherited diseases.[36] For cancer this can be done either by giving the mouse the relevant oncogene which codes for a particular tumour, or knocking out a gene which controls the suppression of a tumour. The resultant mouse is known as an oncomouse.

The aim of genetically engineering the mouse to develop tumours is to provide much greater reliability and repeatability of effect than was hitherto possible. If the experimenter knows that, unless an intervention is made, a certain type of mouse will develop a tumour in a predictable way and at a particular time, then if the onset is delayed or prevented after, say, treating the mouse with a potential therapeutic chemical, then the chemical is having a real effect. To the extent that this can be predicted, the more reliable the test for the potential therapeutic agent, then, it is argued, the smaller the number of mice which will be needed, compared with populations of less predictable non-engineered mice. Another claim is that the greater certainty of the effect allows much smaller quantities of the chemical to be used than in non-transgenic mice, more typical of low dosages to which a human might be exposed. The precision of the modelling, however, varies considerably among the different diseases and genes, but as knowledge based on one type of mouse improves, further mouse models may be developed and refined.

The Harvard Oncomouse as a Human Cancer Model

There are now many different types of oncomouse, but this case study focuses on the first and most famous, the Harvard oncomouse. This not only illustrates many of the issues arising from transgenic disease model mice in general, but it also has special significance to the question of patenting genetically modified animals. The mouse was developed by Leder and Stewart at Harvard University in the early 1980s, substantially funded by the chemical company Du Pont. The patent application describes a variety of ways in which the relevant activated oncogene sequence could be introduced into the early embryo of mice. The result was genetically engineered mice with a much greater predisposition to develop a form of cancer. As it happens, it has apparently been less of a commercial success than many subsequent mouse models.[37]

Patenting the Oncomouse

The Harvard oncomouse has special relevance to patenting, because in 1988 in the US it happened to be the first example of patenting a trans-genic animal.[38,39] The co-inventors filed a patent in 1984, and although other transgenic animal applications had also been filed, this was the first to be granted. This led to severe controversy, in part because the case brought together three serious ethical questions – the acceptability of patenting life forms in general, the particular example of producing a type of mouse specifically to suffer in this way, and the underlying question of genetically engineering an animal at all. The issue of patenting is further considered in Chapter 8.

MAIN ETHICAL ISSUES RAISED

Underlying Issues

The primary underlying issue which the oncomouse raises is the clash between two issues of principle – justification for advances in medicine to alleviate human suffering, and objections to causing deliberate suffering to an animal. It poses serious questions both of animal welfare and of the appropriateness of animal use, especially since the animal has not simply been infected but preprogrammed genetically to develop the disease concerned. On the other hand, it is claimed that the oncomouse is justified by the fact that large numbers of human beings would continue to suffer and die without the benefits that this modelling of human disease would bring. The dilemma is explored in Chapter 5.

Animal Suffering and Human Benefits

What level of human need would or would not justify this sort of intervention and suffering in an animal? To put it the other way round, what level of intervention in an animal would be justified by a given human condition? This introduces the question of the effectiveness of the modelling, the breadth of application of transgenic mouse technology, and the extent to which alternative approaches have been sought.

Commercial and Patenting Issues

The general question is raised, to what extent the motivation for the development of the oncomouse was commercial or humanitarian? Would it have been done if there was no money in it, or would other avenues been pursued? This work also raises doubts of the rightness of making commercial gain from the suffering of an animal. The oncomouse raises the question of whether it is right to extend the scope of patenting to include an entire living organism, or whether this represents an unacceptable commodification of an animal. To what extent could it be considered legitimate to call a mouse, with a life of its own, an invention like a mouse trap? Does transgenesis redefine its status? It also focuses the vexed questions of what role ethical issues should play in the patenting process, and whether, and at what point, the public has an effective say on the ethics of a proposed transgenic development. There are also knock-on effects, in the present way that the system works, such that the granting of a morally debatable patent in one country might create a precedent which virtually forces other countries to follow suit, abandoning ethical scruples under the resulting commercial pressure.

Social Attitudes, Gradualism

The ready availability of disease model mice in laboratory catalogues and the habitual use of expressions such as animal models is thought by some to be symptomatic of a 'DNA ideology' that drives such experiments. Lewontin draws attention to the dangers of an excessive ideological focus on genetics.[40] This could lay open to question the way a society views and respects animals, and whether experimenters see them as tools or creatures. Language and familiarity can legitimise inappropriate attitudes, which could reduce the view of mice by a society to that of mere items in a catalogue. There is also an issue of 'gradualism': having developed a culture which sees the use of disease model mice as a norm, the progressive extension of this could exceed ethical bounds by imperceptible steps.

Case Study 11

Dolly Mixture:
Cloning by Nuclear Transfer to Improve Genetic Engineering in Animals

Ian Wilmut and Donald Bruce

INTRODUCTION

In February 1997, Dolly the cloned sheep became a global news sensation. At the Roslin Institute cells from the mammary gland of an adult had been used to produce a live animal for the first time, using a novel nuclear transfer process.[41] This has in effect rewritten one of the laws of biology. Normally, fertilisation of mammalian eggs is followed by successive cell divisions and progressive differentiation, first into the early embryo and subsequently into all of the cell types that make up the adult animal. Up to now it had been assumed that once animal cells go through this mysterious process, it is irreversible. This work has caused a set of cells to 'forget what type they are' and start all over again, as if they were undifferentiated. While cloning is not strictly genetic engineering, since it copies the genetic complement of the nucleus unchanged, the ability to clone a large mammal from other cells opens up a number of far reaching possibilities in both genetic engineering and conventional animal breeding.

The most immediate significance of this work, and indeed its primary focus, is the possibility of greatly improving the way genetic manipulation is done in animals of the type described in Case Study 8. To date the most commonly used technique is to inject the foreign DNA into an early embryo. This is a very inefficient, hit-and-miss affair, because it has not proved possible either to control where the introduced DNA is integrated or how much of it is integrated. As explained in Chapter 1, as few as 1 per cent of all embryos injected produce a transgenic animal, and even in these the injection does not necessarily have the desired effect. One alternative has been to take embryo stem cells and 'fire' the foreign DNA into them by subjecting the cells to a sudden electric field. These special embryo stem cells have the property of growing into an animal, thus giving a potential of many thousands of transgenic embryos. This is also described in more detail in Chapter 1. Despite a large amount of research, however, embryo stem cell technology has only succeeded to date in two particular strains of mouse.

Nuclear Transfer

The Roslin research has approached the problem in a different way, known as nuclear transfer. This involves two cell types. The nucleus of an unfertilised egg is removed by micro-manipulation. The donor cell is made quiescent, so that the division process is temporarily halted. The donor cell is then placed next to the egg cell whose nucleus has been removed. By passing an electric current the two are fused, and growth resumes. This is shown diagrammatically in Figure 1.5. The resulting embryo is then transferred to the womb of a surrogate ewe and produces what is essentially a clone of the animal from which the cell nucleus was taken. The Roslin Institute transferred nuclei from an established cell line derived from embryo cells of Welsh Mountain sheep and fused these into unfertilised eggs of Scottish Blackface ewes. The result was two genetically identical Welsh Mountain lambs named Megan and Morag, which made headline news in March 1996.[42]

The big step forward in July 1996 (announced in 1997) was in doing the same thing, but starting not from an embryo cell line but from a mammary gland cell to produce Dolly, who is almost genetically identical to the Finn Dorset donor adult ewe which supplied the mammary gland cells. There is a minor difference in that Dolly's mitochondria came from the denucleated donor egg from a different sheep. A key feature appears to have been that the donor nuclei were made to enter a resting state before transfer. Other sheep have also been produced from foetal cells. Evidence emerging shows that a number of cell types may prove to be amenable to such treatments, and it remains to be discovered which are the most suitable. Dolly is not, however, the main line of Roslin's research, because at present foetal cells are the main type being used in further developments of the technique for genetic modification.

Genetic Applications of Nuclear Transfer

The next step was to do the same thing starting with a genetically modified donor nucleus, aiming to produce a transgenic sheep from the modified fused cell. Since it is the transferred nucleus which regulates development, these two techniques together should allow the introduction of precise genetic changes into every cell in all of the resulting offspring, including the germ cells. This was achieved at the Roslin Institute and PPL Therapeutics later in 1997 with the arrival of Polly.[43] The opportunity to target changes in this way offers several advantages. First, it should become possible to use fewer animals in the production of each founder animal because only cells with the desired change are selected for use as donor cells. Secondly, there will be a greatly reduced risk of accidental change to their genes. PPL Therapeutics suggest that for pharmaceutical proteins the method might be used to clone 5 to 10 transgenic animals from a single genetically modified cell culture. These would then be bred naturally, thus becoming the 'founders' of a set of lines of genetically modified animals from whose milk they would extract and purify the relevant protein. There would apparently be no advantage in cloning beyond this first point. Similar methods might be used to produce transgenic pigs for xenotransplantation (Case Study 9).

Perhaps above all, these techniques would also allow changes to be made for the first time to existing genes in large mammals. Until now any changes were restricted to what could be done by introducing new genes. These could for example alter the nature of the protein produced by the cells, the tissue it is produced in, or the stage of development at which it is produced, or even prevent production of it altogether. In research the hope is that this will permit a more precise elucidation of the role of specific gene products and the mechanisms that regulate function of the gene. These are molecular tools analogous to the surgical methods used by physiologists to discover the function of specific organs in earlier research. The significant insights gained into genetic characteristics in animals and also humans could provide new opportunities for the treatment of diseases, as well as the modification of the commercially important characteristics of livestock.*

Cloning in Agriculture

The second area of application of cloning is to conventional animal breeding. Roslin's initial experiments were limited to certain breeds of sheep. Different farm animal species differ quite markedly in their embryology and in the way in which breeding interventions such as artificial insemination and embryo transfer are carried out. The cloning results in sheep are also largely empirical findings. The corresponding theoretical understanding of the processes by which it has occurred will now need to be developed. Much work is going on to see how widely the method could be adopted in other species without adverse effects. In early 1998, similar

---

* The extension of cloning to mice opens up a wide spectrum of medical research applications, since it is much easier to investigate mice than sheep or cattle. Possibilities range from basic research into cell differentiation to the production of cloned human transplant cells from reprogrammed embryo cells. (Wakayama, T et al (1998) 'Full-term development of mice from enucleated oocytes injected with cumulus cell nuclei', *Nature*, vol 394, pp 369–374.)

methods were extended to cattle by two research groups in the US. It is too soon to say how far such developments will be applied, but a number of potential applications in agriculture have been suggested.

In animal breeding, the need to maintain genetic diversity sets practical limits on how far cloning would make sense, but certain applications are already being considered. A commercial breeder of cows or pigs for meat or bulk milk production might wish to clone the most valuable beasts with highly desirable characteristics. These could be cloned to produce lines of breeding stock or to sell the cloned animals to 'finishers' – those farmers who simply feed up the animals for slaughter, rather than breed them to produce more stock. Alternatively, the breeder might want to clone a series of promising animals in a breeding programme, to test how the same genotype responded to different environmental changes.[44]

Conjecture has been voiced of potential applications to racehorse breeding and to the survival of rare breeds, but by far the most important speculative application would be to humans. In view of almost universal ethical opposition, it is unlikely this would ever be attempted in the UK, where it would be illegal under the Human Fertilisation and Embryology Act (1990). It would require an extensive and presumably dangerous experimental programme, but one could not assume that it could never be attempted in other countries with more lax legislation and different value systems.

Main Ethical Issues Raised

Underlying Issues

There is a deep ethical concern that cloning is contrary to something fundamental about life. The very fact that selective breeding has its limits reflects something about the nature of things. Given that biological diversity is one of the features of creation, certainly at the level of higher animals and humans, is the production of genetic replicas on demand intrinsically wrong? In the limit this would mean that cloning would be fundamentally wrong, no matter what it was being used for; or would there be certain applications that would be acceptable?

The Fear of Cloning Humans and its Implications on Animal Research

In the media, by far the most attention concentrated on whether the arrival of Dolly meant that cloning humans was almost inevitable. This was probably premature, in view of the state of the science, let alone the ethical and legal implications. The Roslin Institute and PPL Therapeutics have made it clear that they regard the idea as ethically unacceptable. Legislation can restrict this type of development, but it could never ensure that it does not take place somewhere. It therefore raises a special case of gradualism. Should the research and development of such techniques in animals be restricted because of the danger that they could readily be applied to humans? To what extent is such development driven by the fact that the ability to do it is only a short step in scientific terms, rather than whether it is ethically acceptable? This provides a particular focus for the wider societal question about the way in which research priorities are set, and their accountability to the general public.

Is the Production of Cloned Animals Justified by its Uses?

There are several levels to this question. One is the purely research application of understanding the genetic processes in mammals, and the benefits this may open up in due course. The second is the limited use of cloning to produce a few lines of transgenic animals for pharmaceuticals or other novel applications where there is no 'natural' alternative. The third is the use of cloned mice in human disease modelling and the development of cloned human cells or tissue for transplantation. Finally, there are applications to meat and milk production, where alternative means exist, but would be side-stepped on commercial and production grounds. However logical it might appear from a production point of view, is cloning in routine production one step too far in commodifying nature? Is it right to produce cloned animals just in order to supply supermarkets with meat from the very best animals? These issues are considered in Chapter 5.

Animal Welfare

A number of invasive techniques are involved in the procedure which would need to be justified against the advantages claimed for it. At this early stage, there is also a very low success rate. In the 1996 experiments some birth problems were experienced. Although it is difficult to determine the baseline in a novel technique, some individual animals had much larger than expected birth weights. In 1997, with a different breed combination, the weights were in a better range and fewer deaths occurred. The fact that so many of the cloned individuals die may also indicate that there may be physiological problems. Other individuals appear to be quite normal. The reason for the problems is not understood, but clearly when it occurs, this causes a welfare problem to the animals affected. The different embryological characteristics amongst farm animal species suggest that it is not a foregone conclusion that the methods could be transferred to other mammals without welfare problems. The application of cloning to mice opens the prospect of a significant increase in the number of animals being used in cloning research.

Environmental Effects – Genetic Diversity

There are limitations on how far cloning could be used without violating the need to maintain a sufficient level of genetic diversity in the relevant population of animals. The selected lines could have certain disadvantages in some other genetic trait, which would tend to be evened out in normal genetic diversity of selective breeding, but if animals were cloned, it would tend to cut across this. A limited genetic pool is also much more vulnerable to having all its lines wiped out by a viral or bacterial infection.

3 ETHICS UNDER THE MICROSCOPE

INTRODUCTION

The first part of this book introduced genetic engineering in non-human life forms and illustrated the range of its applications in ten case studies. These were chosen especially to highlight the principal ethical questions which genetic manipulation is raising. The questions are gathered by topic and explored in more detail in the following chapters. Before tackling these more specific issues, however, many of the case studies pointed to certain basic questions which underlie the whole debate over genetic engineering, and these in turn influence many of the more detailed issues. This chapter focuses on these underlying questions, and in particular the intrinsic arguments used by both proponents and opponents of genetic engineering.

Some key ethical concepts are first introduced, leading to a discussion of intrinsic ethical beliefs and their role vis-à-vis consequential approaches. It is often claimed, especially in scientific and official circles, that intrinsic beliefs are out of place in ethical discussions and that the consequential approach is a more rational way of examination. For example, opposition to genetic engineering is frequently said to be based on emotion against a reasoned, scientific advocacy of its beneficial consequences. In the course of the chapter this is seen to be a simplistic view of a much more complex situation. Arguments used both for and against genetic engineering are seen to be related to the different value assumptions found amongst various intellectual traditions in contemporary society, such as the scientific, humanist, Christian and environmentalist.

A number of the more common fundamental ethical questions are then discussed. The first is whether it is right or wrong in principle to do genetic engineering. This is discussed in three ways. The first looks at the notion of 'playing God', either rightly or wrongly. The second examines the idea of unnaturalness, especially by comparison with the selective breeding of animals and plants. The third considers whether or not genetic engineering represents an appropriate intervention in the complex web of relationships within the ecosystem. This leads on to an ethical assessment of the more specific question of transferring genes between unrelated species, and, in turn, to a reflection on human intervention in the chain of organisms from micro-organisms, through plants and animals, to humans, and on the grounds upon which distinctions are made.

The question is posed of how much differing interpretations and viewpoints can be reconciled. Some see them as incommensurable. They represent a wide gulf in contemporary western society between religious, environmental and other groups, who see in the technology of genetic engineering a threat to fundamental values about the nature of life and the environment, and those who take a more instrumental and utilitarian view of it, as the means towards various socially desirable ends. Others consider these differences to be potentially resolvable in plural democracies through public dialogue and discussion. The consultative and regulatory processes are however a barrier in so far as they reflect unequal power relations – between scientific, corporate and governmental interests on the one hand, and the fundamental beliefs of the population at large on the other. One of the main findings of this book is the need for greatly improved mechanisms by which the public can know of and engage in debate about the wider significance of the discoveries and inventions which are being proposed on its behalf.

The last part of the chapter focuses more briefly on the main consequential arguments for and against genetic engineering. These typically involve evaluating the balance of risks and benefits, and this is seen to be an example of a wider phenomenon of contemporary technological societies, in which technologies have created new problems of risk to replace those they set out to solve. Important connections are found between the consequential interpretations of the evidence about genetic engineering and intrinsic beliefs. These interpretations are seen to embody fundamental value systems, both of advocates and opponents, which legitimise these widely divergent interpretations of risks and benefit of genetic engineering. This poses some fundamental questions for society over how such an evaluation should be done, and what a society might and might not legitimately trust.

INTRODUCING ETHICS

Many people regard ethics as something esoteric pursued by philosophers in ivory towers, or perhaps by advisory committees of the great and the good in locked rooms. On the contrary, we all make ethical judgements.

We make value judgements about what we read in the newspapers or watch on the television. When it comes to practical decisions, shopping is one time when we may make many ethical decisions. This might be a choice to buy free range eggs or fairly traded coffee, or not buying wine from a country whose policies we disagree with, or whether it matters to us that something is 'not tested on animals'. Ethical decisions are also expressed in going to the shops by public transport instead of taking the car, the choice of whether we shop on Sunday or take our empty bottles with us for recycling. New shopping decisions we are faced with could include whether to buy food whose production had involved genetic engineering. In medical treatment, we may soon need to consider whether to accept a transplant organ from a transgenic pig or a pharmaceutical product that had come from a genetically modified sheep or a specially engineered plant virus.

These are examples of ethics in practice. There are many ways of making ethical judgements. Sometimes they may be made by plain intuition or out of habit, at others by carefully reasoning. Most such judgements can be classified in one of two ways, described as 'intrinsic' or 'on principle', and 'consequential'. An intrinsic ethical position represents something which is regarded as right or wrong in itself. A consequential one means that it is deemed right or wrong because of the effects or consequences it has. The place of these two approaches has long been the subject of academic discussion, most of which lies beyond the scope of this report. In this study, the validity of both types of argument is recognised, and intrinsic and consequential objections are referred to repeatedly. In practice, the distinction is not always as clearcut as might appear at first sight. Even if some questions about genetic engineering may frequently be posed as if they were simple black and white choices, this book is an exploration of the colours and shades that are found once one probes beneath the surface.

Numerous ethical theories and systems have been developed. Concepts such as natural law, utilitarianism, situation ethics, rights, ecological holism and the social contract have provided frameworks and governing principles within which to work on ethical problems. In this study, however, a quite different approach has been adopted. The deliberately interactive method of the working group discussions has meant a pooling of insights from a variety of ethical approaches. Amongst these, perhaps the most frequent references are made to biblical viewpoints, environmental ethics, risk-benefit assessment and the societal dimension, but the intention is to offer the reader the insights gained from the plurality of our work, drawing conclusions where possible, but more often presenting diverse views on a given issue.

There are also a number of distinctions which it is important to clarify. In popular use, the words ethics and morals tend to be used interchangeably, as though they were synonymous, but there is a difference. The term 'morals' normally denotes the values which are held to be true or important by individuals and societies, such as respect for life, truth telling, and care for the disadvantaged. Ethics, on the other hand, refers to how these are applied to situations encountered in life, and how conflicts between opposing moral values may be tackled. A further aspect of the difference is that although many basic moral values of a society may stay much the same over long periods of time, practical ethical judgements of what is acceptable are more likely to change. This is important for biotechnology since in its current phase of optimism and rapid expansion it needs to recognise that change can go either way. For much of this century it would generally be assumed that change would tend towards more acceptance of new technology, but the last decade has seen a significant shift towards a more sceptical view, which includes criticism of areas hitherto thought to be relatively established, like certain aspects of intensive agriculture.

One factor of note is the phenomenon known as gradualism, in which progressive increments are gradually made in an area of technology, each step being justified on the basis that it represents only a small change from the last. When the overall change is assessed over a considerable period of time, however, it may appear that something unacceptable has happened, when compared with the starting point, without anyone having noticed. A number of examples of this will be noted during this book. The problem is that in focusing only on what the next step represents in its similarity to the last step, sight has been lost of the need to refer each step back to more fundamental values

Another distinction is sometimes made between ethics and aesthetics, where matters of personal taste and preference are seen as different from points of ethical principle or consequence. This is not necessarily so clearcut, however. For example, some argue that genetically modified food is more a matter of taste, in the figurative sense, than ethics, while others would say that sufficient value assumptions underlie our aesthetic sense to make the distinction largely meaningless.

A more important distinction is between ethics and the field of risk, and its associated subjects of human health and safety, and of animal welfare. While they are closely related, safety is not the same as ethical acceptability. Risk assessment certainly involves making important value judgements, but it also includes technical elements beyond the scope of this book, such as what would represent an acceptable health risk. The ethical aspect lies in the value framework against which a risk assessment is made and then presented, and the relative importance a society or group attaches to different types of risks. This is explored at various points in the book, but notably in Chapter 7.

INTRINSIC ETHICAL ARGUMENTS

The Place of Intrinsic Arguments in the Ethical Debate Over Genetic Engineering

In most societies, and until relatively recent times, ethical values and judgements have generally been associated with religious beliefs and cultural practices by which certain actions are held to be intrinsically right or wrong, regardless of their consequences for the welfare of particular individuals or society as a whole. Frequently these were related to sacred texts and teachings. Thus a Christian or a Jew could argue that adultery is always wrong because it is condemned in the Ten Commandments. This reason would be seen as quite separate from an assessment based on the effects of an adulterous act on the individuals involved. A Buddhist might argue that killing people is always wrong, regardless of whether the killing took place in the context of, say, warfare or judicial punishment, because Buddha taught that killing was always wrong. These actions would be considered intrinsically wrong, regardless of whether they had good or bad consequences. Although most scientific and technological advances were not anticipated by the teachings of Christianity or any other traditional religion, an ethical view is possible, with care, by extending their existing principles to these new situations.

Intrinsic views are not only of a religious nature. In some environmental circles there are underlying views that whatever we find in nature untouched by humanity is intrinsically good, and that 'high-tech' developments are inherently suspect. Related examples of intrinsic ethical norms include the wrongness of animal experimentation and of genetic engineering in itself.

For many years there has been a tradition among many scientists and people of a rationalist frame of mind, to dismiss intrinsic ethical arguments as non-rational and emotive. They are seen as of no value or significance in modern societies where rational debate and empirical evidence are the principal canons of authenticity and truth. Intrinsic arguments imply moral absolutes, which are associated either with God or with outmoded philosophical systems, and therefore not to be entertained. Alternatively, they are associated with certain structures in society – groups who have a particular agenda, like the environmental lobby, or perhaps the exploitative power of a social elite. At best, intrinsic arguments are merely one's private view, which cannot be accorded any privileged status, and must be put to one side in debates in a plural society. Since such beliefs are moreover perceived to be unresolvable by rational argument, they ought therefore to be excluded from discussion.

Again, many find a problem with what they see as the 'take it or leave it' nature of intrinsic arguments. If one does not happen to agree with the position put forward, this can lead to problems in finding common ground on which to debate an issue, and especially if there are two opposing principles involved. Some perceive it to be more difficult to engage in meaningful debate with someone who says, for example that to genetically engineer a mouse to be hairless is wrong in principle, than someone who maintains that it is wrong to do so to find a cure for human baldness, because the reason is trivial compared with the suffering to the animal.

For these various reasons, it has long been close to orthodoxy in many influential spheres of government, academia, industry and commerce to dismiss intrinsic views almost entirely from consideration. Yet this apparent expression of liberal tolerance hides the reality of the deep-seated values and belief systems which underlie all our lives, whether consciously or not. Intrinsic ethical arguments are by no means limited to religious believers, or to environmental groups. Indeed, most people hold them in some form. Whether as scientists, journalists, civil servants, company directors, or simply as voters and consumers, most of us hold certain fundamental and often unverifiable beliefs about the cosmos and human life. A notable exception to the above trend was the report of the Ministry of Agriculture, Fisheries and Food committee on emerging animal breeding technologies, chaired by Michael Banner.[45] It is observed that, for example, it is almost universally held that it is intrinsically wrong to torture children. It is typical of such beliefs that they tend to be no more amenable to change as a result of counter evidence and rational argument than is religious belief. The advocates of genetic engineering are likely to be just as much influenced by their own intrinsic beliefs and prior value commitments about the nature of life, humanity and the environment as are its opponents.

Some strongly held beliefs about science also fall into this category. It is maintained surprisingly often that it is only the opponents of genetic engineering who have a value laden perspective towards it. Scientists are claimed to be engaged in a technical procedure which is ethically neutral, and value considerations arise only in judging its applications. This is, however, a serious and short-sighted fallacy. The science of genetic engineering, the interpretation of its scientific data, and the evaluation of its risks and benefits are all laden with the values which are written into the enterprise.[46] For example, the basic assumption that it is appropriate to reduce the physical environment to isolatable genetic, organic and chemical processes, which can be manipulated by the scientist, is a value judgement about the nature of reality and human action within the natural world. Such assumptions as this are so familiar that they are apt to be taken for granted, without recognising that they involve intrinsic ethical judgements. The scientific and technological perspective is as value laden as any other.

Another value assumption is the way in which this application of genetic science is perceived to be of benefit to human welfare. There is a widespread belief that biotechnology provides the answer to endemic problems of the human condition such as poverty and hunger, and will lead to a general advance in human welfare. For some, this can go beyond practical applications and take on the force of an intrinsic and optimistic belief in the power of science, sometimes known as scientism. It is not greatly altered by the somewhat ambiguous record of its application to these questions to date. Those who espouse this view draw attention to the more hopeful evidences, and see past mistakes as things that can be overcome in future. It is closely allied to a notion of 'progress', and has elements of what, in a religious context, would be termed 'faith'. The rightness of the cause of medical science and the eradication of genetic disease are other examples. Although perspectives have changed considerably in recent years, the fundamental belief in the inevitability of scientific progress often remains a barrier to consideration of the wider value questions. It is not only Christians and other religious groups who express a concern to redress this balance, but many others more generally concerned for broader human values, the environment and the developing world.

Thus intrinsic beliefs are not just the prerogative of those who are critical of genetic engineering. Whether positive or negative towards genetic manipulation, to dismiss intrinsic views from discussion, just because they are intrinsic, is to deny something basic, indeed God-given, in human nature. While recognising the difficulties, intrinsic arguments often reveal insights which can throw valuable light on the debate about genetic engineering.

At an institutional level, a radical denial of the validity of intrinsic arguments in ethical debate can present real problems to a democratic process. People approach the arguments about genetic engineering in varying relationships of power and influence, both as insiders or outsiders, with different and often unexposed assumptions and values about the world. In the present context, the danger for those who are in positions of power is to assume that intrinsic beliefs are things which only the opponents of the technology have, and to be unaware of their own, and of the considerable influence that these may already be having. As level a playing field as possible is therefore needed about basic assumptions, in order to avoid the abuse of power. The explicit inclusion of intrinsic questions in the Banner Report was thus a welcome development.

This book, and the working group which has produced it, attempts to redress the balance, in bringing different sets of values into a rational dialogue, reflecting both intrinsic and consequential concerns and views, both favouring and dubious of genetic engineering. By approaching the issues in this way, it is hoped that people on both sides of the debate will be able to understand better the arguments for and against genetic engineering, and especially what underlies them, and so also to appreciate why the debate can become so passionate.

Intrinsic Arguments in Detail

Some of the main intrinsic arguments related to the genetic modification of non-human species are now considered, both for and against. These draw on religious, humanist, environmental and scientific traditions.

The most basic question is whether it is right or wrong in principle to do genetic engineering. The negative view is frequently expressed in popular terminology and by the media that we are 'playing God', or else that it is disturbing something so fundamental to the basic order of nature as to be unnatural. These express two categories of intrinsic concerns about the genetic modification of life on earth. The first reflects beliefs about the divinely given goodness and fitness of life on earth, as experienced in natural evolution. In the view of some theologians, this manifests the wisdom and purposiveness of a beneficent creator, which humans are presumptuous to alter. The second reflects the principles and values of ecological holism, against which genetic engineering is seen as a human intrusion which causes imbalance to the relationships and interactions upon which life on earth is based.

On the other hand, intrinsic arguments can also be made for a proper human intervention in the natural order, both religious and secular. According to these, genetic engineering is justifiable on the basis of notions of human advancement and of care for creation through the practice of human skills in science and technology.

'Playing God' – Positive or Negative?

The expression 'playing God' has been used rather carelessly as a general label for the idea of taking upon ourselves an inappropriate role of changing the way other living organisms are made up. Literally it implies usurping the creative prerogative of God by doing something which belongs to Him alone. Playing, in this sense, denotes taking on a position which was not ours to have. These are matters too high and deep for humanity. It is in God's wisdom that He created all living things and the complex and interlinked means by which they grow and flourish. Human beings cannot match the divine grace and understanding, and our activities must inevitably spoil what God has made. The story of the tower of Babel in the book of Genesis,[47] portraying a massive but ultimately fruitless building project, epitomises the folly of human technological action in autonomy from God.

The genetic manipulation of non-human life is seen as representing a denial of the created or evolved 'good' of animals or plants, and its substitution with human goods and designs. The Bible affirms the independent moral significance of the interests of animals and the limits these interests place on human use. The reduction of animals to manipulable features of the physical environment is a denial of their intrinsic interests. The question of the rights of animals is discussed more fully in Chapter 5. Similarly, the genetic modification of seeds so that plants do not naturally germinate and produce seed for the farmer may be said to involve a denial of the intrinsic fertility of plant life on which hunter gatherers and farmers have for so long relied for sustenance. It may also be seen as a denial of the biblical gifting of plants and their seed to the first man and the first woman in the creation narrative.

It is also suggested that genetic engineering is an act of hubris on the part of human beings, in thinking we can alter the very fundamentals of what God has made. In our human pride we are tampering with something which we do not have the knowledge or wisdom to handle. Many Christians would relate such action to the biblical doctrine that human beings live in a broken relationship with God as a result of rebelling against God and seeking our own autonomy. This loss of our primary relationship to God spoils, in turn, our relationship with everything else – our fellow human beings, animals and nature as a whole. As a result, what can be achieved by human endeavour in science has lost the parallel moral sense of what is *appropriate* to do in nature. We may be so carried away with the pride or wonder of technological achievements that we do not make good judgements about what use we may put them to.

At this point it must be asked why concerns about genetic engineering would place it in such a category, when human intervention in nature is not generally regarded as wrong in Christian thinking. To discern what is appropriate behaviour in this sense first requires an understanding of the way in which human beings are supposed to relate to the natural world. For the Christian, it is more than just nature, or something which has come to be only by impersonal forces of time and chance. Rather, it is creation, the product of the wise, loving and powerful God, and shows the mark of its creator through and through. It is a work of art as much as it is the outworking of the laws of science. It is a gift as much as a system, given to all creatures for their flourishing. It is given to all humankind, but as something to be treasured with wonder as much as a tool to be used. The New Testament declares moreover that the whole created order is held together and sustained in Jesus Christ, the son of God who as the divine Word brought into existence everything that is.[48] He is also seen as the final goal of all creation, in redeeming it from a 'bondage to decay' into which human disobedience has brought it. The natural world is therefore more than merely instrumental to human purposes. It is God's before anything else. A Hebrew poet wrote :

> *The eyes of all look to you, and you give them their food at the proper time.*
> *You open your hand and satisfy the desires of every living thing.*[49]

The creation is ordered to God, from whom it owes its very being.

Human activity in nature is set within this theocentric context. Since it already has another owner and another goal, it cannot be regarded simply as ours to do with as we please. In the book of Genesis men and women are said to be 'made in the image of God'.[50] The meaning of this expression is much debated, but two facets of it are that men and women have both a special relationship with God and also a special responsibility towards the rest of the creation. This responsibility towards nature has been interpreted with a variety of metaphors,[51,52,53] including stewards, guardians, trustees and vice-regents, but, in each case, the world is ultimately seen to belong to God. In the second Genesis account it is expressed in the terms of caring for and working the garden of nature, so that it brings forth its fruits for human consumption.[54] Under the covenant with Noah humans are also given charge over the animals to hunt and kill them for food. Animals are also employed in transport for drawing and carrying, and for sacrifice. As the Old Testament unfolds, human beings engage in a wide variety of technologies of the period – mining for minerals, practising metallurgy, all kinds of craftsmanship, building, irrigating, navigating, and so on.

The biblical accounts thus convey a picture of human beings who are not merely curators of a living museum or rangers of a nature conservation area. God has given the creation and called humanity to understand and to make something of it. The human responsibility for ordering the natural world thus provides the biblical mandate for scientific exploration – exemplified by the deeply significant act of naming the earth and its life forms, which in Hebrew describes the essence of the thing being named. It also gives the basis for the technological manipulation of life – in cultivating and recreating the natural world so that it better serves human welfare, while at the same time caring for our fellow creatures. Another way of describing human activity in nature sees the creation not so much ordered as filled with possibilities, which human beings are to develop, fashioning into new forms the potential of what God first created.[55]

It can therefore be argued that there is a right sense of 'playing God',[56] in which human beings act out God's image before the rest of the created order, in relationship and obedience to God. The Bible recognises many threats to human life within the created world, including famine, flood, drought, disease, earthquake, pestilence and war. A central feature of the narratives of the life of God's people is the ordering of human agriculture, settlement and society to minimise these threats. In as much as it may preserve crops from pests or drought, help prevent famine, and provide humans with healing therapies for disorders and disease, the genetic modification of non-human life could therefore be said to be an intrinsically appropriate use of science and technology for the preservation of human life from the threats which the natural world frequently presents to it.

More positively, scientific discoveries and the ingenuity of technology have served not only to ward off threats but to make advances in the human condition. Christ's command to love one's neighbour as oneself is hereby expressed in helping and enriching the life of others. If science and technology are proper human activities under the God-given mandate to cultivate and reorder the creation, then, it is argued, it is arbitrary to draw a line at the level of genetics rather than chemistry, or at the deployment of molecular genetic engineering rather than the use of fertilisers or pharmaceuticals. Genetic engineering is the latest phase of a technological enhancement of the environment for human benefit. If scientists are prevented from deploying their discoveries, it is argued, the prospects for progressing the human condition will be diminished.

In a humanist perspective, many different rationales may be given for arguing broadly similar conclusions as to the rightness of genetic engineering. For example, human beings are seen to have evolved with unique capacities to use and develop the earth's resources, with natural curiosity, ingenuity and rationality that is expressed in science and technology, and the compassion for our fellow men and women to meet human needs. It is part of human progress to extend our knowledge of the world and our skills in using its resources for human benefit. This goal of progress may be more difficult to achieve than might at first appear, but it is an important human motivation.

These positive views towards the basic idea of genetic engineering stress the potential given to human beings by God, in which we may rightly play God, or the opportunities present in the evolutionary situation in which we find ourselves. The negative sense of playing God is that we take on an inappropriate role in intervening in the natural genetic order. It identifies genetic engineering with crossing over a line of behaviour which is not permitted to us. It challenges an over-optimistic perspective in relation to human capacities and failings, and also for its failure to recognise salient features of the ordering of God's creation which restrain what human beings may do in it. In this latter respect, it has much in common with the ecological perspective which will now be discussed.

Natural or Unnatural? – Genetic Engineering and Selective Breeding

A widespread misgiving about genetic engineering, and to some extent biotechnology in general, is that it is in some way unnatural. Put at its most basic, it is the concern that we are violating something of the given order of the natural world, in making possible by human technology what is in many cases quite impossible among living creatures, or by intervening in a directed way in an otherwise random process.

In ethics the concepts of natural and unnatural are far from straightforward, however. They are difficult to define, because there is a large element of human construction in the way we perceive our surroundings. What is perceived as natural or unnatural inevitably depends on one's assumptions. This is illustrated by the many current and historical examples where both notions have been either romanticised or demonised. We have very different attitudes today about venturing into the mountains, travelling by train or transplanting hearts, compared with, respectively, the middle ages, the early nineteenth century, or a generation ago.

Appeal is often made to some particular concept of naturalness as something given, external to humanity. This is hard to justify on closer examination. It often carries a paradoxical implication that human beings are not themselves part of nature. It would be very difficult to describe any aspect of the biosphere in Europe as natural, after centuries of human activity. In practice, natural may be little more than a synonym for 'how I remember things earlier in my life'. For some, it may just be a matter of unfamiliarity with new technology, as, for example, in some initial reactions to the railways and anaesthetics. The next generation grows up used to the idea and wonders what all the fuss was about.

The natural is also conceptualised to represent some imagined idyllic state of affairs before human beings intervened in the Industrial Revolution, and coming out of the Romantic movement of last century. The numerous examples of the damage earlier societies caused to the environment[57] are testimony that such a pre-industrial 'Eden' has probably never existed, but it expresses an intuition that in all our technological advances we have also lost something vital. The wilderness ideals epitomised by the Sierra Club in the US, or more simply the weekly exodus of thousands of Scottish city dwellers to the mountains, embody something of this intuition. For all its problems, there is nonetheless a significant perception underlying the idea that genetic engineering is unnatural, which is worthy of a closer examination.

What would mark out this technology for such ethical censure? If genes are indeed the most basic building blocks of living organisms, then in manipulating genes directly at the molecular level we are making far-reaching and fundamental changes to other organisms (or indeed ourselves), when we may not have sufficient knowledge or wisdom to do so. Firstly, the general question should be posed, of whether molecular genetic engineering is really any different from the so-called quantitative genetic engineering which we have practised for centuries through selected breeding of animals and plants. Selective breeding can produce effects just as major, or more major, than genetic engineering. The chickens in Figure 3.1 are the same age. The one that looks like a chick comes from a line selected for egg production and when mature will lay eggs for sale. The other is a broiler, selected for growth rate, and is ready to be slaughtered for meat.

At first sight the answer to the question above may seem to be no. Scientists often argue that there is no essential distinction between genetic engineering and selective breeding other than the speed and specificity with which the selection and introduction of genes into new breeds and hybrids can be performed. If we have no ethical objections to hybrid tea roses or to crossing Abyssinian and Ethiopian goats, then what intrinsic grounds are there for objecting to this new sort of genetic manipulation? On closer examination, this is not so simple as a straightforward extension of selective breeding. There are several important points of discontinuity.

The first point is that the things animal breeders are aiming at, like growth rate, lean meat, and disease resistance turn out to be quite difficult to enhance by molecular genetics. To date most of the practical applications to animals are novel ones which could never be done by selective breeding, such as producing pharmaceuticals in milk (Case Study 8), producing organs suitable for transplanting between species (Case Study 9) and using animals as human disease models (Case Study 10).

Photograph: Roslin Institute, reproduced by permission

Figure 3.1 *Two Chickens at Six Weeks of Age*

Genetic manipulation is more readily applicable to plant production, but here again many of the applications are novel ones, which involve genes combined from radically different species. And it is this aspect of *transgenesis* which marks the most obvious discontinuity from selective breeding. Recombinant DNA technology enables us for the first time to mix genes across diverse species, which raises new ethical questions, as will be discussed later.

The second point concerns specificity. Breeding is done by selecting animals or plants which have particularly desirable external traits, like leanness, growth rate and fecundity, which will give an improved yield, or marketable characteristics like a particular colour or shape. Every selection carries with it the complete sum of genetic information of the creatures. In addition to enhancing the intended trait, inevitably other features change at the same time, which may or may not be desired. It may also take several generations to begin to see a significant effect which is aimed for. With molecular genetic engineering, such changes could in principle be done in a single step, with no other genes being altered. It is faster and could be much more specific, provided the genetic change can be made correctly. Because it is more specific, however, this puts a greater stress on getting the modification right. The risk element needs proportionately greater care. Unanticipated side effects may not appear for some generations. If viruses are used to introduce the modification, these need careful monitoring.

The third, more controversial point is that experimenting with the basic elements of life in test tubes is seen as more of a challenge to the natural created order and the laws of nature than conventional breeding methods. The very nature of the procedures involves the reduction of the complex ecological system into its constituent parts which are then manipulated in isolation, or introduced from outside. For some it may pose the question of whether it demonstrates a fundamental lack of reverence for God or for life itself. From an ecological point of view, to what extent is the manipulation being done out of balance with the rest of the animal or plant, or the rest of the species?

The laws of nature have produced a highly complex, diverse and mutually dependent system of communities of life on earth which is well ordered for human flourishing, and for the flourishing of other species. Human beings have for centuries been altering the dynamics of this, but only by the relatively slow steps within the limitations of our ability to direct the course of natural reproduction. Recombinant DNA technology side-steps this by making rapid and highly specific changes. The objection this raises for some is that this side-stepping of the natural processes of breeding involves a denial, or neglect, of inherent processes of disease control and species balance, which may be shown in adverse consequences to biodiversity and agricultural sustainability, particularly in fragile habitats. The consequential aspects of this are discussed later.

Recombinant DNA technology of course raises familiar issues such as animal welfare and unpredictable consequences, but it is frequently the case that old issues assume a fresh significance because of the new context we see them in. People who emphasise that a new technology is really an extension of an existing one with which we are familiar, generally take for granted that the status quo is an ethically acceptable neutral ground from which norm the new developments should be assessed. This begs the serious question of ethical objections to some current conventional breeding practices. For example, the selective enhancement of certain traits –

such as breast or lean meat, or growth – has in some cases proceeded to such a degree that the animal's other normal body functions are significantly impaired. These animals may be said to have been bred to the point that they are now out of balance or that they are experiencing direct suffering. The pressures which are brought as a result of high density, intensive rearing of large numbers of animals also give cause for ethical concern. In some people's eyes, to say that genetic engineering is not inherently different from classical breeding is no great reassurance. It may indeed be greeted with the response that that was exactly what they were afraid of!

In the light of these pre-existing ethical concerns, each new genetic technology should therefore also be assessed in terms of whether it exacerbates or ameliorates them. This point is taken up more fully in Chapters 4 and 5. What would be wrong would be if the undoubted similarities with classical breeding were taken to mean that there is really nothing to worry about over genetic engineering, without properly examining it to see if this was truly the case.

Objections to Genetic Engineering on the Grounds of Relationships

The second aspect of the intrinsic objections to genetic engineering reflects the notion of the relatedness of all life, which is both a feature of the Old Testament and of the principles and values of secular ecological holism. The Genesis account shows two aspects. Human beings in God's image have a dual duty. We are to rule over the rest of creation – 'filling the earth and subduing it'[58] – but we are also to care for it,[59] just as God lavished His own goodness on the creation so we are to do good towards it. This theocentric view of nature implies that neither an anthropocentric nor an ecocentric view are correct frameworks for life on this planet, but the two interests they represent must play together in counterpoint, an ever changing and unfolding dynamic balance. On the one hand, ruling is not meant to imply crushing everything under our feet. We too are creatures of the same God, companions with all created things. We should recognise the relatedness of all of the creation, ourselves included, and act accordingly. On the other hand, this Christian view rejects a notion of ecological perfection in which *Homo sapiens* appears almost as an intruder, and a disruptive influence.

In the view of ecological holism also, there is something intrinsically relational about the diversity of life on earth and its distribution in ecosystems and groups of species. They are environmentally interactive. Different species interact with other species, with climate systems, and water and land habitats in such a way as to sustain the planet and its sub-systems in modes suitable to both human and non-human life. The Gaia hypothesis expresses this in a primarily ecocentric perspective.

The concern in both religious and secular approaches is that with genetic engineering of non-human life we are upsetting the delicate balances that exist throughout the biosphere. We are either imposing our interpretation of what those balances should be, or simply intervening without thought for such effects at all. Again, it should be asked to what extent this would also apply to selective breeding.

The independent moral significance of the interests of animals, and the limits these interests place on human use of animals, is affirmed in the Old Testament, and indeed in many other religious scriptures, and also by secular advocates of the rights of animals. The reduction of animals to manipulable features of the physical environment is a denial of their intrinsic interests. The same case could be argued for plants and micro-organisms, as also God's creations, although with more difficulty, in so far as the Bible and common human practice both accord them lesser moral status than animals. Jesus observes that the lilies of the field are arrayed gloriously today, yet thrown into the fire to be burned tomorrow.[60] Issues relating to animals are explored further in Chapters 4 and 5.

Is it Right to Transfer Genes Between Species?

Molecular genetic engineering has enabled us, for the first time, to intro-duce genes from radically different species – including human beings – into micro-organisms, plants or animals. This is the most dramatic aspect of genetic engineering, from a technical point of view, as it opens up the possibility of adapting living organisms with all kinds of genetic improve-ments taken from other species. Combinations can be created which would be completely impossible in nature or by selective breeding. In one sense, that this is possible at all is evidence of the common genetic heritage of all living things. But it raises a basic ethical question of whether it is right to do something so manifestly artificial. This issue focuses, perhaps more than any other, on whether there is something fundamen-tally wrong with genetic engineering.

As with genetic engineering in general, the arguments in favour are mainly consequential in nature, pointing to the benefits that could result. The main intrinsic argument is that, having given human beings the poten-tial skill and ability to do things which would otherwise be impossible, God always intended that we should eventually reach a stage where we could undertake such intervention, to enable a human ordering of the creation. The creation of new possibilities was part of the way God gave dominion to humankind. To build a bridge or design a wheel made possible things never seen in nature, so then why not transgenesis? Moreover, human abuses of creation have also led to many problems such as food shortage. Might these skills be seen as part of God's grace in enabling there to be solutions even for the results of human folly and selfishness?

The principal argument against transgenesis is that it violates inherent natural barriers among the species and in the given order of nature. Some Christians have understood the Genesis accounts to lay down just such a strict demarcation of species in the description of God making 'everything after its kind'.[61] It is by no means clear, however, that the Old Testament distinction of 'kinds' should necessarily be linked to the biological notion of species. For example, mules are produced by mating horses with donkeys, yet are referred to in the Bible without apparent objection.[62] It is also hard to sustain as an absolute prohibition of the exchange of genetic material as such, since at least at the level of micro-organisms, such exchange occurs routinely in nature,[63] but it might be argued that it could at least apply at the level of higher animals. It is also suggested that deliberate inter-breeding across plant or animal species was specifically forbidden in the Old Testament law.[64] It is a moot point, however, whether this falls into the category of ceremonial law, symbolising in this case Israel's distinctiveness, which for the Christian has been superseded through the death and resurrection of Jesus Christ.

It is valid at this point to ask how important is the concept of species. It is an attempt to rationalise and classify the biological diversity of the natural world. This classification is a human construct, but it recognises similarities and differences that genuinely exist. In that sense, species represent an aspect of the natural order. How far should human adaptation seek to override this? It could be argued that we should not go as far as to alter radically aspects which are seen as inherent in the order itself. The question is then whether *barriers* between species are something inherent to this order. In terms of evolutionary biology, some might argue that the idea of barriers is artificial. We have a continuum of genetic possibilities which is grouped into discrete nodes at any one moment in time, but it is in a state of continual, if slow, variation. In mammals a great deal of our genetic material is common. Yet however much DNA they actually have in common, pigs and humans, for example, are clearly different. They have developed down different branches of the genetic tree, and the differences are actually rather significant, both in their biological effect and upon our perceptions. If small genetic changes can have such large effects, we should therefore be careful what we are doing when mixing genes across different branches since we cannot know the full implications. But it would be difficult to maintain on mechanistic grounds that species as such constitute an absolute prohibition.

A stronger argument concerns wisdom, balance and a much broader sense of ordering in nature. Christians might argue the biblical passages cited above indicate a wider moral teaching which is still relevant. Even if the letter of the law is superseded, there is something in the spirit of it.[65] There is a wisdom in the natural order of things which reflects the goodness and purposiveness of the creator. For humans to mix aspects of different organisms by genetic engineering would go beyond God's wise ordering of life. Humans are not given to have the degree of responsibility for life on earth in this way. We have neither the omniscience to anticipate all the dangers, nor the right to exercise such power over nature. It would actually usurp the responsibility of God. It is therefore another example of inappropriately playing God.

The same argument could be made from an environmental point of view. The differentiation and balance of species which has come about through evolution is something intrinsically valuable. There is a natural wisdom about it, and we should not override it. If natural processes have resulted in a world suited to the flourishing of humans and an enormous and wondrous diversity of non-human life, it is argued that the distinction of species and orders should be respected and not altered. What they are in themselves is of intrinsic value, so we should not seek to dilute the distinctions by mixing genes. To insert foreign genes into particular species, and then introduce the transgenic species into the wild, would represent an assault on the laws of nature, as well as possible consequential effects of endangering the health and diversity of ecosystems. Intrinsic objections of this kind would also apply to the cell fusion experiment that produced the sheep-goat chimaera known as a geep.

The idea of the natural (or divine) wisdom of the natural order is an important principle. The fact that we can accomplish transgenesis which is impossible in nature is clearly something profound, and not something to be done lightly. To argue that it should not be done at all clearly has some logical as well as intuitive basis, and not a few people hold such a position. It raises, however, the question of why the line of non-intervention should be drawn at this particular point. Chimaeras, or some other ways of mixing large numbers of genes between species, do indeed raise serious doubts about the level of appropriate intervention in the ordering of species. It is harder to maintain, however, that changing one or two genes constitutes a wholesale violation of something we should not be changing. A sheep with a pig gene is still primarily a sheep. The dividing line may fall over the question whether the true nature of something is in the detail or the wider essence. For those who see it in detail, transgenesis will be unacceptable, as a change to a blueprint. For those who see it in the essence, then the limited scale of genetic alteration represented by some of the case studies in this book would not constitute crossing over a line.

Even if, as most of the working group would do, one adopts the latter position, the insights of these intrinsic concerns should not be lost. They undoubtedly demand that a considerable measure of care should be taken in mixing genes between species, aware that there is a balance and integrity to the genetic make-up of the organism which receives the foreign genes. The drive of a technical mind set to engineer novel constructs, to see if they work and can be exploited, should not overwhelm the sense of respect for some of the givens of creation.

Before leaving the issue of transgenesis, it is important to consider the special case of the genetic intermingling of human and animal DNA, which could raise special problems. There are again various Old Testament prohibitions,[66] some of which are clearly ceremonial regulations and others moral law. More broadly, some vital features of human moral and social life flow from the distinction we draw between humans and non-humans. Human life is uniquely valued, and the killing of humans is forbidden in virtually all ethical systems except as an extreme punishment or in time of war. It might be argued that the routine use of human genes in animals, or vice versa, could undermine the moral claims we associate with human life. This in turn depends on the relative status we accord to animals, plants and micro-organisms in relation to ourselves as human beings.

The Status of Animals, Plants and Micro-organisms

The ethics of doing genetic manipulations in different species is significantly affected by how, and in what sense, distinctions are made in the chain of being from micro-organism to humans. It is a commonplace that we do not regard all living things as of comparable status. Broadly speaking, the 'higher' the organism the more value humans assign to it. Consequently we consider we have less right to intervene. Higher tends to mean more like us, for example more intelligent, complex or sensitive. We have more reluctance to intervene with apes and large mammals, than small mammals or fish, and in decreasing degrees insects and other invertebrates, plants and micro-organisms. An emotive, but no less real, response is that we are more concerned about causing harm to something that is warm and furry than if it is cold and slimy, or microscopic.

A number of ethical approaches attempt a more rigorous approach by focusing on the criterion of sentience – the ability to be conscious of pain and other stimuli. This is one of the most important criteria used to differentiate what we may and may not do to our fellow creatures. It was argued in the case study on *Pseudomonas* (Case Study 1) that no one is seriously worried about harming or abusing a microbe as such. Our main concern is not about hurting it, but rather whether it could get out of control and do harm to humans. A similar perception is made towards most plants, insects and fish, with the additional concern for conserving habitats and species. It is only when we get to large plants like trees, or to animals – and particularly mammals – that major qualms tend to appear. When asked if it was better, all other things being equal, to obtain proteins for medicinal use via the genetic manipulation of plants or animals (Case Studies 4 and 8), many would choose plants as the invasion is seen as less important. Although plants certainly make responses to stimuli, this does not appear to equate with the pain and discomfort that mammals can exhibit.

There is a strong pull towards increasing the status we accord to animals, especially as our knowledge grows of their comparative similarities with humans. However, there are dangers in taking this too far. An evaluation made at one point in the chain has a knock-on effect. At one end, it has been observed by some that the closer one moves animals, in concept, towards humans (so increasing one's duty of care to animals), the more ready is the danger of treating some humans merely 'like animals'. Thus if sentience were the primary criterion there would logically be an argument for experimenting on the mentally disabled or those in a persistent vegetative state. On the other end of the chain, it can be said that the further one moves animals towards humans, the further one moves them away from plants and micro-organisms. At the present point in time, when the consequences of a general lack of respect for the environment have become all too apparent, this would not be a trend to recommend.

It is significant that it was a micro-organism which was the subject of the ground-breaking decision in US patent law which has opened the door to a general policy to allow patents on living organisms. In effect, the Supreme Court judgement evaluated the status of the micro-organism by extension from biochemistry, as though it was scarcely more than that of a chemical. Whereas this might be plausible at the level of the micro-organism, subsequent patent practice does not appear to have stopped to ask whether there is any difference in status between microbes and mice. Having taken an initial analogy with chemicals, patent law *de facto* extended the analogy to all life forms, more or less without distinction. This is discussed further in Chapter 8.

There is clearly a difficulty in finding a scheme which conveys the distinction of respect which we feel intuitively belongs to the chain of being, without favouring certain parts unduly at the expense of others.

Any functional approach such as sentience diminishes the deeper question of the inherent worth of all creation. At this point a Christian understanding sheds valuable light. If every organism is itself a product of God's generous and bountiful creation, it is therefore of intrinsic worth, without regard to its sentience, its function, or its utility, or otherwise, to humanity. This is not to say that, as in a pantheistic view, God's nature dwells in them, as such. Rather, in being what they are, they reflect and glorify God. Because God is not embodied in them, this means that humans can use them without doing irreverence to God, but we cannot use them in any way that we like without disrespect to the creature. The Bible drops many hints which imply distinctions in the relationships among the orders of creation, but it does not lay down a definitive scheme by which to delineate which orders one may or may not modify genetically. A possible clue, however, may be that the sense in which human beings feel they can relate to another creature may offer some guidance to what is permissible when it comes to genetic engineering.

Summary of Intrinsic Ethical Viewpoints

The discussions in this section show a wide range of possible viewpoints by which either genetic engineering as a whole, or at least aspects of it, might be seen as intrinsically right or wrong. From a Christian perspective, the abrogation of the order of creation which God has laid down is seen by some as a bar to such modification, especially in transferring genes among different species. Other Christians regard God's command to humanity to 'fill the earth and subdue it' to contain a mandate to explore and utilise the earth from which genetic engineering could not be excluded on principle, yet his parallel command to care for the creation sets a restraint, that not all that could be done should be done. The notion of playing God can be seen in a positive as well as a negative light.

From a more secular stance, some would see it as intrinsically right to use the potential of the earth and its creatures to forward the life of humankind as our skills enable us, and to combat disease – both human and animal. To describe genetic engineering as 'unnatural' in an absolute sense has been seen to raise considerable problems. The perspectives of ecological holism, reflected also in some feminist and Christian approaches, pose a significant objection, however. Genetic engineering could be seen as upsetting the balance and relationality of nature, and the orders and diversity found within it. The new powers opened up by molecular genetics go beyond the restraints of selective breeding. The fact that the latter has manifested some serious abuses towards animals is for many a pointer that we have gone far enough already.

The role of a Christian critique is especially important. Not only does it represent the community of the church, as possibly the largest identifiable body of public opinion in the UK, but it articulates underlying concerns common to a much wider circle of people. Those expressing worries about globalisation, biotechnology and technology in general without any particular frame of reference on which to pin them have often found many echoes in the application of the Bible's ancient insights of the nature of reality.

But how are we to deal with these intrinsic pros and cons? Scientific and bureaucratic responses to intrinsic objections to genetic modification tend to be couched in the terms of consequential arguments, cost-benefit and risk assessment, and overlook the intrinsic nature of the objections. In Chapter 10 a sociological critique is presented of the surprisingly prevalent assumption that science is objective and intrinsic objections are merely emotive and irrational. It is important therefore to repeat that intrinsic beliefs and values are reflected on both sides of the argument. They are present in the scientific community's advocacy of genetic engineering as well as the objectors' opposition to it. It is important to recognise the fundamental views of the world that underlie them. We tend to be more familiar with such views expressed in opposition to genetic engineering, but the ethical rightness of any research, for example, towards curing cancer or enabling the world's population to be fed are examples of deep-seated views expressed in support of it.

CONSEQUENTIAL ETHICAL ARGUMENTS

The discussion so far has concentrated on some of the intrinsic issues which underlie genetic engineering, to set the background for examining the more specific areas in the subsequent chapters, where the focus will shift somewhat, but not exclusively, to more consequential issues. Before embarking on this, some general points should be made about the consequential approach to ethics:

- It is the most widely used in modern societies.
- It seeks to weigh the sum total of the likely benefits and disbenefits represented by a particular course of action to those likely to be affected by it.
- No prior judgement is made as to what is a good or a bad action in principle.
- The notion of 'good' is seen as a pragmatic and procedural one.

- It is typically expressed as the greatest welfare to human persons, when it is generally known as utilitarianism.

In the present work, however, the scope is extended to include consideration of the effects on the wellbeing of non-human life and the environment, and not just humans. This recognises that the harms which can result from human actions towards the environment or non-human life may ultimately also prove to be disbenefits to human welfare.

Potential Benefits of Genetic Engineering

Chapter 1 listed a range of both potential and actual applications of genetic engineering, expressed primarily in technical terms. A wide range of potential ethical benefits could be cited in support, as the positive consequences which it is argued would justify such developments. These include:

- feeding an increasing world population;
- economic benefits to poor regions by making marginal agriculture more viable;
- reducing environmental degradation from agriculture by reducing chemical inputs and more efficient land management and production;
- reducing other forms of environmental degradation by novel genetic applications in pollution control and alternatives to fossil fuels;
- medical benefits – improved understanding of human disease and developing improved therapies by novel uses of genetics in animals, plants and micro-organisms;
- veterinary benefits – understanding and treating animal diseases;
- production benefits – reducing inputs and overheads in many stages of agricultural production;
- consumer benefits – including food quality and nutrition, and wider choices;
- wealth and job creation;
- intellectual and cultural benefits to the society from excellence in genetic research.

To these the general question may be asked: which of these potential benefits would serve to justify genetic engineering in the creatures and organisms with whom we share the planet, and which would not?

Potential Drawbacks to Genetic Engineering

There are a comparable range of negative consequences of genetic engineering, which include:

- animal welfare problems, both in the experimental and production stages;
- encouraging the treatment of animals more as manipulable commodities or bioreactors, and less as creatures in their own right;
- the risk of transferring disease from animals to humans, or antibiotic or herbicide resistance to unwanted organisms;
- the uncertainties involved in releasing into the environment new life forms which cross species barriers – their effect on wild varieties, weeds and pests, and the overall ecological balance;
- reducing genetic diversity by relying on a smaller pool of genetically optimised products;
- enhancing food surpluses in richer countries, while dumping on developing economies;
- disparities in the distribution of both the benefits and disadvantages of genetic engineering;
- disadvantaging those who object to genetically modified food;
- lack of public accountability in the developments;
- the skewing of research and applications towards Western consumer markets, so that wider world needs are neglected;
- the focus of agricultural science on genetic engineering perhaps representing a distraction from research into more sustainable forms of agriculture which might more effectively reduce some of the more damaging effects of modern intensive farming on soil, water, biodiversity or climate.

Some General Observations About Consequential Issues

Each of these is explored in more detail in the relevant chapters, but several overall points may be made about consequential ethical assessment. Firstly, there can be a significant disparity between the optimistic potential which is claimed for a new scientific breakthrough and the real benefits which accrue in production. There is often a tendency to overstate the possible benefits both in the heady enthusiasm surrounding such a breakthrough, and also in the more pragmatic phase of putting the best case forward to attract investors. The way the potential is portrayed in the media is often an overstatement of the case. Myths can develop which may appear to justify going ahead, but which can prove difficult to live with later if problems begin to occur, or if the project does not deliver as many benefits as the investment prospectus had anticipated. For example, a primary justification put forward for genetically engineered model mice – that they would reduce the number of mice used in experimentation – has now been brought into question (see Chapter 5), as has the need for BST in a time of milk surplus. An ethical evaluation of the benefits of novel genetic technology needs a measure of both humility and realism. On the other hand, some of the risks which are claimed can be exaggerated or multiplied without limit, with the intent of never allowing anything to go ahead.

This raises again the need to recognise the different value judgements which underlie whatever are cited as the principal benefits or objections. The ethical goods of scientific progress, human medicine, enhanced agricultural production, consumer benefits, or reduced environmental degradation all imply certain intrinsic values about what is important in life. These are not self-evident givens, but arise from particular worldviews.

The same is true of the fears that are expressed. It is important to appreciate this when these benefits are placed alongside the disadvantages which might result from the application, or the alternative advantages which would be gained if other objectives were sought. For instance, in Chapter 5, fundamental ethical views can be seen to play a significant role in weighing the relative merits of human medical benefit and animal intervention. Taken to the extreme, either medical advance on the one hand or animal suffering on the other may even be regarded as justifications strong enough in themselves, and thereby take on something close to an intrinsic ethical status. When opponents and advocates of genetic engineering adopt consequential arguments for or against it, they do so in part because of their intrinsic beliefs about such things as science, justice, or the status of animals, and not purely on the basis of arguments from consequence or risk assessment. In considering the positions adopted in debates in this area, utilitarian views may thus be rather less important than these deeply held intrinsic beliefs and value commitments, which are sustained by particular communities and traditions, on both sides of the argument. Each has its community – for example, represented on the one hand by protest against modernity, or on the other by the scientific, corporate and technocratic belief in the capacity of modernisation to deliver health, prosperity and justice.

The need for a sound economic case in order for a genetic application to proceed may skew the ethical consequences quite strongly. An application with a sound ethical justification, such as enhancing food supply within poorer countries, may be much disadvantaged by comparison with products whose main good may merely be Western consumer preferences.

Weighing Up the Risks and Benefits of Genetic Engineering

How do we weigh up such a complex series of pros and cons? The moral framework in which our society generally operates is at root utilitarian, based on assessing the balance of risks and benefits. The term cost-benefit analysis is often used to describe this, but is open to misinterpretation due to its monetary connotations. Its origins are in numerical process in which the factors involved are evaluated, and balanced according to their economic worth. Ethical assessment can never be quantitative in this sense. Cost-benefit is sometimes used, in contrast to intrinsic ethics, to imply that the consequential approach is rather more scientific and rational than the process of informed value judgement which it really is. The economic associations of the idea of cost-benefit analysis can also convey the notion, associated with certain trends in economic and political philosophy, that human value questions can ultimately be reduced to monetary ones. It is thus not helpful to use terminology suggestive of monetary evaluation when discussing things that are, in most people's terms, beyond mere quantification, like the prevention of starvation, therapies for terminal disease, loss of genetic diversity or animal suffering. If a term is needed, then risk-benefit analysis is perhaps preferable, especially since consequential objections to genetic engineering are often expressed in terms of the risks of this new technology – for example, to human health, animal welfare, the environment, or to society in general.

The idea of 'weighing up' is also numerical, but it is a very qualitative weighing, which is not concerned with measuring whether something is 0.1 gram over, but asks more broadly whether something feels enough or too much. There are at least three types of weighing up, which might be expressed as:

- 'Yes/no' (0 or 1, 100 per cent or zero, on or off), a question of state. Is it acceptable or not to modify pigs to provide heart transplants for humans?
- 'How much' – simply evaluating the quantity, not questioning the state. Does the oncomouse involve too much animal suffering?
- 'What balance' – how does it compare with the competing claims of other things? Are genetically modified tomatoes justified when there are serious food shortages in developing countries?

In a complex society, most questions are ones of evaluation. Comparatively few questions are truly in the either/or category, but there will always be some in any society or belief system. Indeed, it is the challenges to these that most unsettle people.

Underlying the Discussion of Risks and Benefits

It is striking how we have changed what threats we are most concerned to protect ourselves from. In pre-modern cultures the fear was primarily from naturally originating hazards such as storms, wild animals, droughts, floods, and famine. Within the Christian framework, these threats were understood in the context of our need for divine protection and sustenance – well reflected in many of Thomas Cranmer's prayers in the Book of Common Prayer. In the modern world, the threats that occupy our minds are almost all of human origin. The environmental risk from technologically originated hazards – such as road accidents, pollution of land, air and water, nuclear melt-downs, and food contamination – has become the new horizon of threat from which modern people seek to order and protect their lives and those of their children. This is a major factor in, and expression of, the secularisation of modern society in which religious belief and its influence have moved more to the margins of European society. There is a deep irony in this. In so far as people have believed that the scientific explanation dispenses with the need for religious assurances about one set of threats, this has created another set of threats from our very 'overcoming' of nature. Many Christians now argue the need for the *wisdom* of a religious perspective to handle such questions. We are never free of threats; we simply shift them.

The shift in what threatens us may also represent a subtle change in the way the human self relates to the world and to nature, whose cultural and environmental implications are still unravelling. The resistance to new technologies such as genetic engineering in some quarters may be partly a function of the uncertainty of this process and its implications. The unfettered optimism of the technicist view of the world is clearly no longer satisfactory for many, perhaps most, people. It is equally clear, however, that we cannot simply reject science and technology outright, as if to return to some imagined pre-industrial Eden.

Far from being something which technology has freed us from, the weighing up of risk has thus become a characteristic feature of modern societies. Ulrich Beck's term the 'risk society'[67] struck a chord in describing the extent to which modern societies are prepared to countenance great risks for the progression of economic growth and human welfare.

Like many others with strong environmental concerns, he is doubtful that it is good for the quality of life and environment in the long run for modern societies to rely on a risk mentality to the extent that we do. This is discussed further in Chapter 10. An important philosophical point is that he indicates a prior scepticism about the intrinsic scientistic belief that a technologically altered environment is necessarily one which will promote a general and quantifiable improvement in human welfare and justice. He is not alone in this. Other notable examples are Mary Midgley[68] and Egbert Schuurman, who has made a similar critique of what he calls technicism, on Christian presuppositions.[69]

Beck also places a considerable emphasis on the capacity of the power relationships in society to impose more of the risks of new technologies on marginal or powerless groups. The elites gain most of the benefits in the new balance of natural resources or welfare enhancements which new technologies may create. Because of the way society and human nature runs, there are likely to be losers as well as winners from new technology. In a world of different ethical values, the key processes of societal decision-making are on the one hand the expression of power in markets and on the other private and publicly expressed passion and protest. The rise of single issue social movements in relation to environmental and other ethical dilemmas is a reflection of the growing tension between power and protest in late modernity.

Grounds for Assurance?

On what grounds should a society be reassured about the emerging genetic technology, having once seen cause to express doubts and fears about it? Risk tends to be something which experts assess and then tell the public about. But the public looks for certain things in order to have confidence in what it hears. A long track record of good practice is certainly one prerequisite, but this is not enough in itself. As well as good technical performance there needs to be the right attitude and philosophy, and this is in many ways more difficult. For some, there is an intrinsic distrust of the science or the scientists, which may not be answerable to their satisfaction, or perhaps only after a very long time. Many need considerable reassurance that responsible approaches have been taken with respect to what are termed the 'unknown' consequences of genetic engineering, given what problems have come to light in so many other areas of new technology. In the limit, this would be a counsel of unattainable perfection, but pleas for genetics to proceed more slowly clearly have a point. For many others, whose concerns are more specific or consequential, there is a need to see signs that responsible caution is being practised by all the different players – research scientists, the bioindustry, governments, regulators and ethical committees. In order for there to be a justifiable sense that these key players acting on our behalf are in control of what they are doing, a sense both of limits and of humility has to be conveyed. There should be an appreciation of what it is like to sit in another's shoes and hear how one's carefully worded public statements actually come over to people.

Genetic engineering could do well to take several leaves from the pages of the saga of nuclear power. The claims and reassurances made in the past for nuclear power are an object lesson in the dangers of exaggeration or over confidence. A sense of humility needs to be conveyed on the part of the proponents of genetic engineering, so that they do not claim more than is fair, or assert what they do not know. Given that things go wrong with most technologies, there also needs to be a sense that mistakes have been learned from, not only on the specifics of the causes of an event, but the underlying attitudes and structures which contributed to it. Just as the nuclear safety watchdog needed to be seen to bite for the public to be sure it was doing its job properly, so there needs to be an assurance that there are some things in the area of genetic engineering and biotechnology in general beyond which we would not go. The cloning of Dolly has given a graphic illustration, where one breakthrough drew a very widespread public expression of the need to declare boundaries and to say that *not* all benefits of biotechnology are desirable.

Incommensurability is not Inevitable

Taken to the extreme, it might be said that there would be no way of adjudicating between the radically opposed arguments for and against genetic engineering which have been laid out above. Ethical incommensurability is said to be an inevitable feature of a plural society. The coexistence of many different beliefs and traditions is a characteristic cultural condition of late modern societies. It could be argued that in such a society intrinsic views, strictly speaking, have ethical power only within the value systems of identifiable groups. The relevant groups here include Christians, other faith communities, scientists, the biotechnology industry, animal campaigners and environmentalists.

As the group producing this book has demonstrated, however, there is often considerably more scope for dialogue between supposedly incommensurable views than might be thought possible. Strongly held positions either way may be re-evaluated in the light of fresh angles, or by realising hitherto unappreciated implications. What becomes more difficult is when an issue becomes strongly politicised. When wider political interests line up on one side or the other, it polarises the debate into implacably opposed positions, and after a time it becomes almost impossible for one side to hear the other, as with the case of nuclear energy.

Genetic engineering has not reached such a stage, yet, but it is already pertinent to consider how a society copes with what Oliver Wendell Holmes called 'a people of fundamentally differing views'.[70] If these views are minority ones, it is a fundamental principle of democracy that minorities should not presume the right to impose their views on majorities. If the majority consider that their welfare may be enhanced by research and deployment of GMOs, then ethical minorities cannot expect to stand in the way. Equally, the majority needs to respect a minority view, give it fair hearing, and to avoid as far as possible forcing those who hold it to violate their ethical principles. This is especially relevant to the question of labelling genetically modified foods (see Chapter 6).

Public Participation in True Dialogue

Care is needed for incommensurability over genetics not to result in violent social conflict of the kind reflected for example in the actions of some pro-life and ecological campaigns in North America. Greater efforts need to be made by the public bodies and corporations who fund and sustain the scientific community's genetic engineering projects, to promote genuine dialogue between objectors and advocates of this new technology. It is not, however, sufficient to engage in public relations campaigns or the somewhat patronising idea of 'educating the public'. This mistake has been a sad feature of many responses from bioindustry leaders, government spokespeople and the European Commission. The implication is that the traffic is largely one way. It is assumed that the experts will educate the public in what is what, and once they understand they will accept the technology. In its most extreme form, bioethics is reduced, as some have said, to the function of lubricating oil whereby the initial friction caused by new scientific discoveries is overcome, once the public come to understand that their fears were groundless.

While not denying the importance of scientists explaining what they are doing, and clearing up misconceptions, of which there are often many, experience has shown that a patronising attitude does not work. There must be an explicit expectation of two way communication. For science to be truly accountable, its proponents must give the public leave to say 'We have heard you but we are unconvinced about certain developments', and to affirm its right in a democratic society to reject some new technology if it so wishes. Indeed, one of the key needs is for working scientists to be exposed to perspectives outside their normal professional circles, to answer for their work in an environment which may not reinforce their own outlook. Our own working group has illustrated the mutual value of doing this, but it has also shown that it requires a commitment to working together over a reasonable period of time. This small group model might be appropriate for setting up an ongoing engagement between the scientists of a research institute and a humanities faculty of a nearby university. A different sort of approach is the notion of the Consensus Conference. This is a national or regional event in which the scientific community is subjected to public scrutiny by a lay panel in a two way exchange. There is probably no universal approach. Much work is clearly going to be needed to develop a portfolio of approaches to address this vital dimension of accountability of biotechnology.

Advocates of genetic engineering argue that the dilemma of incommensurability is resolvable in plural democracies through various means of public dialogue and discussion. It must be said that this claim is somewhat called into question, when it comes to translating debate into policy-making, if it does not also recognise the social construction of power relations in modern nation states. Opponents of genetic engineering argue that consultative procedures and regulatory processes simply reflect the unequal power relations between scientific, corporate and bureaucratic stakeholders and opponents of biotechnology. Moreover the principal actors in these processes and structures have thus far tended to share a fundamental belief in the inevitability of scientific progress and in the inherent connections between scientific progress and advances in human welfare.

This uneven playing field has rendered these procedures open to suspicion, and suggested that there is a problem of democratic accountability in the development of biotechnology in the UK. The balance of membership of various official regulatory, safety, advisory and ethical committees needs to be widened considerably, to reflect a greater diversity of view than has hitherto been the case. It is not sufficient to have the token 'green' or 'consumer' person if the bulk of a committee is dominated by industry representatives. This may satisfy the letter of the law, for the purposes of answering a parliamentary question, but it misses the spirit of it. Such discussions tend to hear evidence and then do everything else in private, without any further intermediate discussion with the public. This is not real dialogue. The model for such assessment needs revising. More satisfactory approaches appear to exist in, for example, The Netherlands national committee for transgenic animals, where the provisional findings of the committee are opened to public comment before the final recommendation is made to the government minister. At the local level too, procedures need to be established whereby communities adjacent to both experimental and production sites are kept informed and consulted.

CONCLUSIONS ON UNDERLYING ISSUES

This chapter has examined the two main approaches to assessing the ethics of genetic engineering, and found much of value and insight in the intrinsic ethical questions alongside the more familiar consequential evaluation. As the different issues are discussed in the succeeding chapters, both will be used, as appropriate. While some will undoubtedly find reasons to reject it, a good case can be made that genetic engineering need not be regarded as wrong in principle, whether from Christian, secular or environmental perspectives, but there is also much to be concerned about over its implications. There is a context of technological development in both agriculture and medicine that has already aroused much concern. It should not be assumed that a point would never come when, on balance, a society should decide it had gone far enough with high technology in this area, when wider considerations were taken into account. There is as much to criticise in an unfettered ideology of scientific progress as there is in a fundamental rejection of all genetic technology.

A consistent theme is how the fundamental beliefs of a society and the different players have a key role in how genetic engineering issues are interpreted, both intrinsic and consequential. Chapter 10 will return to this theme, adding a sociological dimension. Genetic engineering has come to fruition in a period where the trend has been increasingly to questioning previous generations' confidence in science. It has a task made all the harder by the unexpected difficulties of some other modern technologies. How a plural society deals with this in a just and equitable manner is not straightforward, but some obstacles have been identified. The remainder of the book will explore in detail how particular issues may be examined, seeking to throw light on complex questions as a contribution to the ongoing process of ethical evaluation of genetic engineering.

4 GENETIC ENGINEERING AND ANIMAL WELFARE

INTRODUCTION

The previous chapter considered both consequentialist and intrinsic approaches to the ethics of genetic engineering and the interaction between these. Consequential issues include effects on human welfare, animal welfare and the environment, and the present chapter will consider in detail the implications of genetic engineering for animal welfare.

Most ethical evaluations in this area have considered the broader field of biotechnology rather than just genetic engineering. Two particularly systematic evaluations may illustrate this: a workshop on 'Biotechnology and Animal Welfare' financed by the Organisation for Economic Cooperation and Development (OECD)[71] and a committee set up by the UK government to consider 'Ethical Implications of Emerging Technologies in the Breeding of Farm Animals' (The Banner Committee).[72] The committee listed a number of techniques beyond the scope of the present study – artificial insemination, superovulation and synchronisation of oestrus, embryo transfer, *in vitro* fertilisation and semen and embryo sexing – as well as cloning and genetic modification. There are two points to make about the drawing of boundaries. First, although we are concerned here with the impact of the relatively new technique of genetic engineering we should not assume that the status quo is an ethically acceptable or neutral ground. Indeed, there is no status quo. Selective breeding continues with most of the animals with which we are concerned, and welfare problems have resulted and continue to result from the increasingly sophisticated methods used. Examples of such welfare issues in farm animals are the physical problems associated with rapid growth in broilers and turkeys, the hunger and frustration of feeding behaviour caused by food restriction of broiler parents and sows and the calving difficulties of double-muscled cattle. The ethics of such 'traditional' methods are increasingly called into question. Indeed, ethical questions are also raised about whether animals should be kept (and in many cases killed) for human use at all, but this issue will not be discussed here.

Second, however, concern over new breeding technologies is primarily caused by the fact that effects on welfare may be more rapid and intense than hitherto. In particular, genetic modification has vastly more potential than other techniques for producing sudden change in characteristics relevant to welfare. It is more akin to mutation than to recombination of genes already present. For this reason, this chapter will concentrate on genetic modification of animals and only consider other techniques where necessary. Where relevant it will also mention modification of hormones and vaccines which are then used on animals. Further references may be found in the reviews cited, including that by Appleby.[73]

EFFECTS OF TECHNIQUES USED IN ACHIEVING GENETIC MODIFICATION

The techniques used in genetic engineering often cause welfare problems irrespective of the nature of the modification achieved. Some of these problems are associated with husbandry and are similar to those of any other manipulative experimental work. They can be divided into three categories which act separately and in combination. First, human contact may be frightening, for example during handling or due to changes in the predictability of husbandry routines. Second, social conditions are often stressful: isolation is common. Lastly, physical conditions are usually suboptimal, with barren surroundings and diets which can be consumed rapidly. In some countries such problems are being considered more than hitherto under the general licensing procedures for experimental work, but there remains considerable room for improvement. One promising finding, with potential to reduce the ill-effects of forcible restraint in many species, is that pigs and sheep can be trained to enter a restraining device for procedures including blood withdrawal (and hence potentially also for injection of anaesthetic) voluntarily and repeatedly.

Other techniques include cloning to create and copy GMOs and producing embryos by transferring the nucleus of one cell into another without a nucleus. Embryos resulting from this nuclear transfer have a high rate of mortality and at least some have been unusually large (see Case Study 11), so that there may be welfare problems for both the offspring and the mother around the time of birth.

The actual techniques of genetic engineering may involve a number of procedures on a number of animals. For example, the following welfare issues are associated with pronuclear injection in mice: hormone injection (to stimulate superovulation), mating (while still only three to four weeks old) and euthanasia (to obtain the fertilised eggs, into which DNA is then injected) of mothers, sterilisation of males (which mate with foster mothers to produce pseudopregnancy), insertion of embryos into foster mothers (done under anaesthetic), mortality of offspring and ear punching or tail tip removal of offspring (to distinguish transgenics).

With some applications, techniques are used only a limited number of times, because once genetically modified lines are established they breed true. This is so, for example, with farm animals used for biomedical products. With other applications, for example research into embryonic development using transgenic mutants, techniques are used repeatedly. With such routine techniques it is important that the precise methods and their implications for welfare should not be taken for granted. It should also be remembered for all applications that there will be welfare problems for animals kept in reserve but not used and those on which the techniques are unsuccessful, as well as for any transgenics produced.

For some applications, in which the changes produced are themselves neutral for welfare, the sort of effects considered in this section may be the most important welfare consideration. None of these effects is specific to genetic engineering, so they have received relatively little attention and perhaps some people have felt that they are not a cause for special concern. However, any such consideration must be made against a background in which the justification of all animal experimentation is increasingly questioned, with pressure for the 'Three Rs' of reduction, refinement and replacement.

EFFECTS OF GENETIC MODIFICATION

If genetic modification is successful it will result in changes to the physiology and perhaps the physical structure of the animal concerned. Many of these will have direct implications for welfare, intended or unintended. Modification is also likely to have indirect effects, with the animal being treated in ways which are different to normal husbandry. This section will consider the types of effects which are possible and the following section will discuss the extent to which they actually occur.

Direct Effects

No Effects

Some modifications appear so far to have no direct implications for welfare. An example is transgenic sheep such as Tracy which have the gene for human alpha–1–antitrypsin (AAT) and produce this protein in the milk (Case Study 8). No ill effects are apparent, although checks on this are still proceeding.

Planned Effects

Some changes may be described as specifically intended to affect welfare, either in a positive or negative way. One positive effect which has been attempted in a number of studies is increased disease resistance in farm animals. We may also mention here genetic engineering of antibiotics and vaccines to improve their efficacy and specificity. Benefits to welfare could potentially also include measures to reduce or prevent other specific problems such as leg disorders in turkeys or the stress of close confinement in sows. However, genetic engineering is expensive so the only improvements to welfare which are likely to be implemented on a voluntary basis are those which are profitable. The ethics of this commercial approach are considered below.

Intentional negative effects on welfare exist where transgenics are used to study disease. This is most often justified by their use as models for similar conditions in humans, and the appropriateness of the model is a major issue here. The approach could perhaps also be used to develop treatments for animals, but this does not appear to have been done to any extent. Animal suffering is an integral aspect of this work, as will be discussed below. Mice are most commonly used, for example the oncomouse with increased susceptibility to cancer (Case Study 10), but livestock species such as pigs are also studied.

Side Effects

These have probably received more attention than any others. Perhaps the best known example is the Beltsville transgenic pig,[74] with enhanced growth hormone production, which had very severe arthritis. Similarly, sheep with additional growth hormone genes never attained puberty and died before they were one year old. Robinson and McEvoy[75] state that 'In many instances, the site and time of expression of the transferred genes still lack the degree of specificity required and lead to deleterious side effects'. Loew[76] makes the trenchant point that in new work 'Unanticipated results, like those … described in swine by Pursel et al,[77] are, as it were, to be anticipated.' In the case of Loew's specific example, however, the results of Pursel et al should have been expected: it was already known that high concentrations of growth hormone in pigs (achieved by injection) caused liver and kidney degeneration, oedema, arthritis and leg disorders. In a number of applications, work continues to reduce such side effects.

Some effects arise as side effects but then become subjects for study, as with the Legless mouse, considered in the following category.

Integral Effects of Other Changes

Some modifications have effects on welfare which are not themselves the reason for making the modification but are integral to it and so can not be considered side effects. The distinction is ethically relevant because if such effects are inherently unavoidable they must be accepted and justified as an inevitable consequence of the modification. Transgenics in which normal immunology or development is disrupted, studied for the insight they provide into normal processes, come into this category. Thus the Legless mouse has major limb and craniofacial abnormalities and dies within 24 hours of birth, and is used in embryological and genetic research.[78] This category would also apply if production of pharmaceuticals in the milk of farm animals had deleterious effects on those animals or if genetic engineering increased production in ways with similar effects to those mentioned in the Introduction for selective breeding (such as rapid growth and double muscling).

Indirect Effects

Husbandry and Related Effects

As an indirect result of genetic engineering, many animals are kept or treated in ways which have other advantages or disadvantages for welfare. Transgenics are valuable and their health tends to be looked after particularly well. However, as with other experimental animals, they are often kept in isolation rather than in more natural social groups and the importance of hygiene usually means that they are kept in barren conditions. Indeed, for some purposes they are delivered by caesarean section and kept in sterile or near sterile conditions to keep disease risk to a minimum. By contrast, one concern about the work on disease resistance is that if it is achieved it may make it possible to stock animals at higher densities; this possibility also applies to animals treated with vaccines improved by genetic engineering. Isolation, barren conditions and high stocking density all cause problems for welfare. These are similar to the problems which may be associated with the experimental stages of genetic engineering, discussed above, but in this case may be applied permanently rather than temporarily.

Another indirect effect, resulting from the genetically engineered improvement of growth hormones, is increased frequency of injections. Thus if BST is used to enhance milk production of dairy cows it usually has to be injected daily (Case Study 7). In addition to restraint of the animals and the pain of the injections themselves, there are also problems with injection site abscesses. Some implants are used, for example lasting two weeks, but these have to be injected with a thick needle which is probably more painful. Work on longer lasting implants continues.

Numbers of animals kept for different uses will change as a consequence of genetic engineering. Some uses may be more efficient and need fewer animals: this is one of the aims of work on growth rate in farm animals. Other uses increase numbers; thus the number of animals used in transgenic research is currently increasing rapidly. There is no simple correlation between the numbers of animals involved and the importance of their welfare, but there is probably a consensus that some association does exist. The overall effect of genetic engineering on welfare will therefore depend on the balance between problems and benefits for the animals' welfare (including problems and benefits which are independent of the genetic modification) and the increase or decrease in their numbers. The three Rs of reduction, refinement and replacement should again be borne in mind here.

Effects on Attitudes

Modification of animals is likely to affect attitudes to them and hence other aspects of their treatment; it may also alter attitudes to and treatment of other groups of animals. This applies both to people who have direct influence over animals (such as breeders and producers) and to the public, whose influence is nebulous and rarely focused. Attitudes may be affected by new uses of animals (for example as models for human disease or as suppliers of organs for xenotransplantation) and by changes in their legal status (such as whether particular types of animal can be patented). Implications for welfare are difficult to predict: the effects of attitudes to animals are complex.

CATEGORIES OF ANIMALS

Farm Animals Used for Agricultural Products

The main attempts to date to change production characteristics of farm animals by genetic engineering have been insertion of genes for growth hormone into pigs and sheep. There is also similar work on fish. It has been suggested that one possible benefit of biotechnology would be more efficient production leading to the use of fewer animals. In fact there seems to have been little consideration of whether such animals growing faster or further would actually be more efficient economically, in terms of food conversion. In addition, transgenes would need to improve economic performance by 5 to 10 per cent to be useful, because they would take several generations to introduce and meanwhile the performance of other stock could be improved by normal breeding methods. However, use of fewer animals for the same meat production would seem to be an advantage. So far, though, most of these attempts have had gross side effects as described above and it is clear that such modifications will not be used commercially unless these can be avoided. Investigations are continuing, for example into whether it is possible to have the gene present but 'silent' during early development, then expressed later, which should prevent the worst problems. It has been pointed out that in elderly humans injection of growth hormone may have positive effects, increasing lean body mass, decreasing fat and increasing bone density.

Cloning by nuclear transfer may be used to increase the specificity and range of genetic modification in breeding stock. It may also be used, as described in Case Study 11, to copy particularly productive animals. This might seem to give a relative advantage for welfare, because although cloning is associated with welfare problems these are probably less than those of other currently available procedures such as manipulation of growth hormone. However, this does not mean that it is necessarily justifiable. Furthermore, improvements in other lines will continue, so that such clones are unlikely to be the most productive animals for very long.

A cause for concern is the possibility that future work will produce changes with commercial advantage and with side effects which are less obviously unacceptable – similar to those which have been produced by selective breeding – because there will be commercial pressure for such deleterious side effects to be tolerated. This is the situation with the use of BST to enhance milk production of dairy cows, which is practised in the US but currently banned in the EU. This has limited effects on welfare if management is good, but negative effects if management is poor. In some respects it is therefore not a special case, but comparable to other management techniques (such as housing design and feeding regimes) intended to increase the profitability of production. However, it could nevertheless be seen as unwarranted in its application of technology.

Some changes in production characteristics being investigated or sought are likely to be neutral or positive for welfare. Work in The Netherlands has produced Herman the transgenic bull, whose female progeny are intended to produce milk containing the human protein lactoferrin. This would make it more digestible for babies and patients on antibiotics. It is believed that this change will also reduce the risk of mastitis in the cows. Another area of interest is the possibility of producing hens and cows which only have female offspring – for egg production and milk production respectively – avoiding the necessity to kill male chicks or rear unwanted male calves. The unnatural character of such developments concerns some people, but they may nevertheless reduce the incidence of welfare problems.

Optimism about the prospects of increasing disease resistance in farm animals appears to have abated recently. Some of the approaches being investigated were unsuccessful or restricted to very specific experimental circumstances. In addition the technology mostly concerns single genes, whereas the disease organisms concerned are complex, and it is increasingly recognised that any increase in resistance might only be temporary. In many cases a slight change in the organisms (for example, by mutation) would be sufficient to make them infective again. It is true that there are examples of long term resistance of certain species to certain diseases, for example native African cattle to trypanosomiasis (sleeping sickness). This suggests that if it becomes possible in future to change multiple genes rather than just single genes some permanent change in disease resistance might be achieved. However, there are again other ethical issues here.

First, in relation specifically to trypanosomiasis, it has been suggested that the only reason there are still large areas of Africa relatively well populated with wildlife is the incomplete resistance of cattle to this disease. If resistance is increased or a vaccine is developed this will greatly increase the pressure on such areas and make conservation of wild animals much more difficult. Second, some of the diseases prevalent in current production systems have been exacerbated by intensive selection for production and by the techniques used in those systems, for example mastitis in dairy cows. Unless genetic modification of dairy cows for resistance to mastitis reduced incidence of the disease to what it was before the increased production, it could be argued that it would be more appropriate to reverse the changes which have caused the problem. This is particularly true if there are other ill effects of increased production on welfare which these techniques help to perpetuate. Similarly, modifying the animals to prevent disease may increase the tendency to keep animals in poor conditions (such as high stocking density, as mentioned above) which have other disadvantages for welfare.

Farm Animals Used for Biomedical Products

Of all applications of genetic engineering, this one currently has most commercial potential. As with animal production, the welfare issues are not wholly new: some farm animals are already used for biomedical products with welfare problems resulting. For example in North America many thousands of mares are kept in stalls too small for them to turn round, for the production of oestrogen from their urine. Transgenic animals do at least tend to be kept in conditions which ensure their health, although with other limitations. These, and the welfare implications of the techniques used in genetic modification, were discussed earlier.

The area of work which has received most attention is modification of sheep or goats to produce pharmaceuticals in their milk for human medical use (Case Study 8) which will be cheaper and safer than those from alternative sources (such as human blood). The changes being made – or at least, those being publicised – appear to be neutral for welfare. As Loew[79] asks, 'What possible harm to man or beast can arise from a minor change in the composition of goat's milk such that it becomes a cost-effective source of a valuable pharmaceutical?' Yet vigilance is necessary, because certain genes might not be expressed solely in the mammary gland, and because the milk–blood barrier is not complete, so some compounds will be expected to affect the lactating female. For example, there has been interest in use of this approach to obtain erythropoietin, which regulates erythrocyte production and can be used to treat renal disease. Yet human or monkey erythropoietin has severe or fatal effects in mice when it circulates in the blood, and even if expression of the regulating gene is restricted to lactation it can not be assumed that the compound will not leak back to the blood. A cow transgenic for the human erythropoietin gene has been produced in Finland but has never been allowed to produce milk and hence allow the gene to be expressed, partly for this reason. Similar considerations apply as for farm animals used for production: gross effects on welfare will be unacceptable, but there will be commercial pressure for toleration of more minor side effects.

One other area which is receiving increasing publicity is the modification of pigs to allow their organs – heart, kidney or pancreas – to be transplanted into humans, as discussed in Case Study 9. One approach being tried is insertion of a gene for human complement regulators into the pig genome, which will label the surface of pig cells so that hyperacute rejection does not occur when they are transplanted. There is no reason to believe that the welfare of a pig with such a gene will be compromised during its life, but it may be necessary to keep the animals in barren environments. The other ethical issues involved with xenotransplantation are discussed in the following chapter.

As with farm animals used for agricultural products, cloning may be used to copy animals which are particularly appropriate for pharmaceutical production or xenotransplantation. Again, this is likely to involve welfare problems, but these may be less than those caused by repetition of the procedures which are otherwise necessary to produce such animals.

Laboratory Animals

Most lines of research are still at the exploratory stage, so in that sense all animals involved are laboratory animals, but the term is used here to mean animals – primarily mice – on which work is being done without immediate application in that species.

Many procedures carried out on laboratory animals are disturbing. As indicated above, some involve intentional production of major welfare problems, as in the oncomouse, and others require tolerance of such problems, as with the Legless mouse. However, the issues are clearly complex: one point that was made in the public discussion of the oncomouse (in relation to whether or not it could be patented) was that in a particular study use of such a strain would make it possible to use fewer experimental animals. On the other hand the increasing availability of transgenics is increasing the number of experiments being carried out and the number of experimental animals which are suffering. Figure 4.1 shows the increasing use of transgenic animals in the UK between 1991 and 1996. In these statistics from the UK Home Office they are included in a category of 'animals with harmful genetic defects' produced by both genetic engineering and other methods.

The intention behind production of such animals is largely philanthropic – prevention or cure of human diseases – although there is necessarily variation in the applicability of such work. Thus Cameron et al[80] list the following areas of investigation: gene regulation and development, host-pathogen interactions, immunology research and oncology research. Of these, the first is probably further from application than the others, yet mice studied for this reason probably suffer just as much. This becomes relevant when regulatory bodies decide whether particular

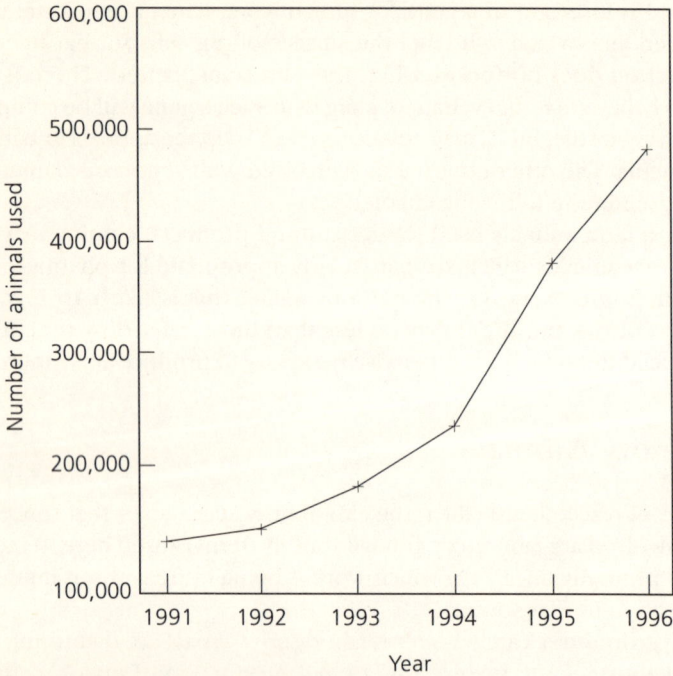

Source: Home Office Statistics from Stationery Office, 51 Nine Elms Lane, London, SW8 5DR

Figure 4.1 *The Increasing Production of Transgenic Animals*

research proposals are to be allowed. Laws in the UK, for example, require consideration of both the likely effects on the animals involved and the likely benefits. However, assessment of the balance between these is clearly difficult. Poole[81] has pointed out that animal models of human diseases are sometimes inappropriate: even if the symptoms are similar they may have very different causes and development, and may also exist in combination with other pathologies which do not occur in humans.

The issue of patenting has mostly been concerned with this group of animals. Patenting of a whole animal (such as the oncomouse) is permitted in the US but is still being debated in Europe, while patenting of DNA sequences is permitted in both areas. The issue is important here because patenting may affect welfare. Its effects are likely to be complex, but one confusion should be avoided: patenting does not in itself confer any right to use animals or to condone suffering (these are regulated in the normal way). One point which has been made, though, is that experiments on animals which are restricted by patenting are likely to be done in reputable laboratories, while those on animals which are not so restricted may be carried out in conditions which are less than ideal.

It is likely that more transgenic animals are being produced and are suffering than are needed even for the experiments being carried out, because laboratory animal breeders often keep animals in stock to anticipate demand. It is clear that animals should be bred only for firm orders rather than always being available. Genetic modification of mice to make them prone to disease causes suffering, often severe. The ethics of this will be considered in the next chapter.

Other Animals

Little or no work is being done on genetic modification of sporting animals, companion animals, zoo animals, wild animals or pests. Perhaps there is potential for improvements in disease resistance in the first four of these groups, but as discussed for farm animals, that does not currently seem likely.

Genetic engineering is being applied to vaccines and viruses to be used in wild animals which are seen as dangerous or inconvenient to humans. Recombinant vaccines against rabies are being used in foxes, racoons and skunks and are beneficial to those species insofar as they reduce other control measures directed against them. On the other hand genetic engineering is being used to introduce an immunocontraceptive effect to viruses which will then induce sterility in pests such as foxes and rabbits. It is intended to use the myxoma virus for rabbits, which will limit the disease caused to the rabbits because many are now immune to myxomatosis and will simply become sterile. The question of pest control is, of course, a complex one: many issues such as conservation are involved as well as human inconvenience, but it still appropriate to point out that lack of concern over the welfare of pests is inconsistent with strong concern for the welfare of other animals. In this case there will also be other issues such as the risk of these viruses affecting animals other than the intended ones.

EVALUATION

One of the issues arising from the discussion so far is the question of why the work is being done at all. For many applications it is clear that a major driving force behind genetic engineering is commercial exploitation of technology. This is important here because of the complex interaction of economics and welfare, which makes it unlikely that any applications will be unequivocally beneficial to both animals and humans.

If there is to be proper evaluation of the justification for genetic engineering, what will be the contributions of the general public? One contribution, which affects the commercial underpinning of the work, is the willingness of individual people to use the products of the processes we are discussing. People's willingness to buy food from genetically modified animals will be affected by various factors including the group of animals concerned: for example, genetically modified fish may be more acceptable than mammals or birds. Perception of the animals' welfare will also be important and will be affected by the information available. For example, with limited information people may have deep but rather formless concerns about the potential effects of genetic engineering on animal welfare, whereas with full consideration it is at least possible that those concerns will be fewer. There may be an initial impression of dozens of grossly suffering animals like the Beltsville pig going into commercial production which is not borne out by the facts. Nevertheless, it seems reasonable that increased safeguards are necessary.

The other way in which members of the public continue to have an important influence is in relation to such safeguards, because safeguards are put in place by legislators and legislators are influenced by public opinion. In North America and in many countries of Europe, members of the public who are concerned about animal welfare are more vocal than those who are not – joining humane societies, writing letters to the newspapers and expressing strong views in opinion polls. This climate of opinion increases the pressure for ethical evaluation and for the control and legislation of genetic engineering.

Many discussions and conferences on genetic engineering include consideration of ethics, particularly about implications for animal welfare, but such articles often make no practical recommendations. One article which does do so is that by Mepham,[82] on the basis of potential advantages and disadvantages to humans and animals. He provides a framework for cost-benefit analysis of the effects on the animals and on the different groups of people affected by the technology (farmers, consumers, people in less developed countries and so on). This is described in Appendix 3. As with all cost-benefit analysis, this does not provide quantitative answers but does clarify the relevant questions. In particular, the question recurs of how to assess effects on animals. Perhaps the most practical recommendations in this respect are those by Broom.[83] With reference to farm animals he suggests that 'carefully controlled studies using a wide range of welfare indicators are needed. These should be carried out for at least the total farm life of a breeding animal and for at least two generations.'

Broom goes on to point out that 'No such comprehensive studies ... have been reported in the scientific literature to date'. However, there have been concerted welfare assessments made on a lesser scale, eg Hughes et al[84] compared transgenic and non-transgenic sheep. The question remains as to whether the public can have confidence that such assessments will be applied wherever appropriate.

LEGISLATION AND CONTROL

It might be argued that few of the welfare implications discussed above are different in kind from those of other procedures such as selective breeding and hence that additional legislation is not needed. However, against this it must be pointed out that current legislation and welfare codes are not tackling existing problems successfully. For example, at the 1994 European Symposium on Poultry Welfare it was concluded that

> *The growth rate of broilers and the concomitant food restriction of broiler breeders cause major welfare problems. ... These problems are getting worse with the continuing, intensive genetic selection for growth rate. ... Urgent consideration should be given to legislation against further selection for growth rate until or unless associated problems are solved.*[85]

The inadequacy of current legislation is also likely to be worse with the potentially more rapid or unforeseen effects of other breeding technologies.

For adequate control of such technologies, legislation is necessary which specifies what is permitted rather than what is not. This is particularly necessary for procedures involving genetic modification. To obtain such permission, the proponents of a procedure should have to demonstrate one of two cases. First, the procedure may have no deleterious effects on animal welfare. Alternatively, benefits to humans may outweigh deleterious effects to animals.[86] Arguments for the latter case must be assessed by a properly constituted process such as that used by the UK Home Office under the Animals (Scientific Procedures) Act 1986 for licensing animal operations. Without such rigorous control, it seems likely that genetic engineering will cause more disadvantages than advantages for animal welfare.

The argument in the previous paragraph is currently tautologous in the UK, because all genetic engineering must be done under the Act mentioned. The UK also has regulations on potential release of GMOs, monitored by the Health and Safety Executive. However, not all countries have similar legislation. It is also unclear in the UK whether some procedures will in due course come to be regarded as routine rather than experimental and thus be excluded from the requirements of the Act. Another limitation of this legislation is its low profile. There is little public accountability, and yet it is essential for public confidence in the safeguarding of animal welfare that the procedures of committees concerned with these issues should be well publicised. The Banner Committee[87] stated the position with regard to farm animals well when it said:

> *One unlooked for consequence of the introduction of the emerging technologies can quite reasonably be anticipated – and that is the creation of public suspicion of farming, unless those who are engaged in the development and application of these technologies endeavour to be sensitive to public concerns, open to debate with interested parties and supportive of a reasonable system of regulation, provision of information and labelling.*

It went on to recommend 'that an advisory committee be created, whose remit should include a responsibility for broad ethical questions relating to current and future developments in the use of animals.' Such a committee was described as 'an ethical watchdog' in the press;[88] one already exists in several countries such as The Netherlands[89] and it is important that other countries including the UK should establish committees of this nature or similar mechanisms, or that this should happen on an international basis. Furthermore, it is essential for public confidence in the safeguarding of animal welfare that the procedures of such committees should be well publicised.

Conclusions

The application of genetic engineering to animals may affect their welfare in a number of ways. There may be direct effects of the intended change, side effects and indirect effects. Some modifications are beneficial for welfare, some are neutral, and some are detrimental for the animal. It is of note that few genetic applications of direct benefit to animals have been attempted and they have not proceeded beyond an experimental stage. For example, although various theoretical possibilities for disease control exist, the disease organisms concerned are genetically complex, and any increase in resistance might only be temporary.

In considering the welfare impact of genetic engineering, it should not be assumed that the status quo is an ethically acceptable or neutral ground. The ethics of traditional breeding methods are increasingly called into question. Against this background, genetic modification in animals has few effects on welfare which could not also be produced by selective breeding, but it has vastly more potential than other breeding techniques for producing sudden change in characteristics relevant to welfare. Moreover, where it enables the development of techniques which would be performed routinely on animals, the acceptability of the welfare implications should not be taken for granted.

At present, the main applications are not, as originally expected, in the field of enhanced animal production, but in novel modifications for biomedical purposes. The three important examples considered are: the production of pharmaceutical proteins in the milk of farm animals, which can, in the best examples, be largely neutral for welfare; xenotransplantation, which has greater but not necessarily overwhelming welfare implications; and the modification of mice as models for human disease, which results in suffering of large numbers of animals. Their ethical implications are discussed in the following chapter. As work in the field of animal applications progresses, the situation is likely to change rapidly and needs to be kept under constant review, especially should animal production techniques become possible.

In no country does current legislation avoid welfare problems from selective breeding. Additional safeguards are needed for the rapid and repeated changes which genetic engineering can effect, which should include controls on the use of recombinant hormones, vaccines and viruses in animals. Procedures are in place in some countries to evaluate the impact of genetic engineering on welfare, but these need to be strengthened. Where they do not exist they should be established.

One of the driving forces behind genetic engineering is that of commercial exploitation. There is concern that this will put pressure for deleterious side effects to be tolerated. An example is the use of the recombinant hormone BST in dairy cattle, which can have negative effects on welfare unless management is good. Some of our group regard this as an unwarranted application of technology.

The commercial aspect also raises concerns over public accountability. Ethical evaluation of work in this field is still in general done on a case-by-case basis, with limited public involvement, and using the limited criteria seen as directly relevant to each case rather than a broader framework. For the public to have confidence that such evaluation is being properly carried out, a wider approach is needed, with much greater public participation.

Calls for advisory 'watchdog' committees to consider ethical questions on the use of animals are endorsed here. Furthermore, it is essential for public confidence in the safeguarding of animal welfare that the procedures of such committees should be well publicised.

5 ANIMAL ETHICS AND HUMAN BENEFIT

INTRODUCTION

The previous chapter examined what implications genetic engineering could have for animal welfare. A number of problem areas were identified, against a context of our present understanding of what is entailed in the good welfare and well being of our fellow creatures. This chapter now assesses how far these negative aspects may, or may not, be justifiable on ethical grounds. It looks at them both in absolute terms and in the context of the benefits that are sought. As was noted in Chapter 1, most applications of animal transgenesis to date are targeted not at conventional improvements in animal production, but at novel medical applications. Hence this chapter focuses in particular on the dilemmas posed by animal intervention and suffering for the purpose of alleviating human suffering through medicine, in respect of case studies 8, 9 and 10 on transgenic animals – pharmaceutical proteins in sheep's milk, xenotransplantation and mouse models – and Case Study 11 on animal cloning.

This chapter is thus an exploration of how the potential of genetic engineering affects the underlying ethical question of how far human beings are justified in intervening in the lives of animals for our own benefit. Before going into the detail of the case studies, a brief review is given of how this issue has been seen by some of the existing ethical frameworks, and some alternative ideas are suggested.

APPROACHES TO ANIMAL ETHICS

Throughout human history, animals have provided humanity with many things including food, clothing, traction and companionship. In almost every society the use of animals is accepted, with widely varying degrees of reservation and limitation. Technology, which has made some of these traditional uses redundant, is now coming up with other ways of using the creatures with whom we share the planet, which pose new challenges. The primary argument put in favour of genetic engineering of animals is that since we already use animals in such a variety of ways, and especially that we kill them for food, why should we not also manipulate them genetically for human benefit? The main constraint is seen as the need to ensure that animal welfare issues are properly attended to. Having done this, some would maintain that there are no further ethical issues to be addressed.[90] The fact that an animal is transgenic is seen as largely irrelevant ethically.

Against this rationale, a wide variety of perspectives have been articulated in the field of animal ethics, which seek to define the relationship between animals and humans, beyond simply questions of suffering or welfare.[91] In the extreme case that it was thought wrong in principle to use animals for any human use, and more specifically to take an animal's life, then even saving a human life by using an animal would be ruled out. None of the following outlooks espouse so radical a position, but many of them would strongly question whether genetic engineering is an appropriate use of animals because of the significance and status of animals, on a variety of grounds. The importance of considering intrinsic issues of this type has also been underlined by both the Banner Committee[92] and Reiss and Straughan.[93]

A Survey of Animal Rights and other Approaches

The most familiar of these perspectives is perhaps that of animal rights, which was first proposed by Tom Regan.[94] He espouses the view that all mammalian life has inherent value because of their experience of sentience – consisting of pleasure, joy, pain, identity, memory, a sense of the future and potential for fulfilment. He terms mammals as 'subjects of a life' and as such regards them as having the same rights as humans who suffer when treated immorally but are not themselves morally responsible – for example infants and the mentally ill. On this basis he argues that animals have rights and that the duties owed to animals are almost the same as those owed to humans. Thus it is as morally wrong to hunt or rear animals for food production, use animals in laboratory experiments and eat meat, as it would be to do these things to humans. This would rule out genetic engineering on principle.

This ethical position draws much on the similarities between humans and animals in sharing the property of life. Mary Midgley[95] criticises this view by pointing out that giving rights to some animals tends to mean that we give less to other animals. If sentience is taken as the criterion for rights, how then are the needs of different beings like locusts, tapeworms and human beings to be balanced? In addition, the rights framework does not give us any basis for caring for trees and plants. She also argues that it is not necessarily wrong to treat different species in a different manner, since they often have different needs, and indeed it would be wrong for us to ignore these differences. Her view is that preferring our own species is a natural part of our existence. Nonetheless, just as sometimes we subordinate our family or group interests for that of outsiders, we should be prepared to extend this principle to animals. She also points out that we are often apparently inconsistent in our views. For example, we may treat our pets almost as proto-humans and yet be happy to eat meat which has involved killing an animal.

Andrew Linzey[96] argues that the key to the recognition of the independent moral value of the creation lies in the doctrine of God as the generous creator who values all that is created. He proposes somewhat unusually that all creatures are Spirit-filled individuals and we can thus affirm that they have rights, which he has termed 'theos' rights. The argument is that God as creator has rights in His creation, so animals can elicit moral claims which are nothing less than God's claim upon us. This sets limits to what humans may do to change the intrinsic nature and integrity of other sentient beings. Again, this would preclude genetic modification. Our ideas of human good are not independent of what is good for other sentient beings. In later writings,[97] he emphasises the Christian concept of the higher sacrificing itself to the lower, as a model for what is implied when the Bible speaks of human dominion[98] over animals. He argues that the principle of generosity is in fact stronger than the principle of rights. His main emphasis is again on individual sentient animals.

Criticising this position, Barclay[99] maintains that the fact that God has rights in His creation is an entirely different situation to animals having rights. Rights only make sense in human affairs when seen alongside responsibilities. Since animals cannot have comparable responsibilities, the language of rights is inappropriate to apply to them. It is better to talk of human duties towards God, regarding animals.

Peter Singer takes a somewhat different approach by extending the consequentialist, utilitarian framework to include animals.[100] The aim of obtaining the greatest good for the greatest number should be extended beyond merely human good to include the moral interests of animals, because they too are sentient. He coined the term 'speciesism' to refer to a prejudice against animals which he sees exhibited in human society. In this utilitarian framework, there may be occasions when the life of an animal has priority over the life of a human. An example would be the case of killing a young pig to use its heart valve in a transplant operation for a senile person. Logically, Singer's position would imply that preference would be given to the pig, since animals and humans are, at a given level of consciousness, to be treated as equals. By using sentience as a yardstick in this version of utilitarianism, again, no basis is offered for valuing non-sentient life, let alone other non-living aspects of the world around us. Singer's approach also runs into an opposite danger of cheapening the life of human beings, in a world where animals are put to death or subjected to great pain for trivial reasons such as cosmetic testing.

Environmental ethicists have tried to avoid the problem of drawing lines between animals and the rest of nature. Holmes Rolston[101] uses the concept of intrinsic value, based on the independent existence and purposiveness (the *telos*) of an organism. This allows an objective moral value to be given not just to individual animals, but also to ecosystems, species and the whole biosphere. Rolston has also developed a hierarchy of values by which the different interests of organisms and systems may be discerned. Michael Fox[102] uses this approach in arguing that genetic engineering alters the *telos* or 'beingness' of an animal as well as its relationship with the external environment. Considering respect and reverence for life, the integrity of the species and the sanctity of the individual, to produce genetically engineered organisms for human profit is therefore unacceptable.

The Dutch philosopher Henk Verhoog argues that each animal and plant has a good of its own which brings them into the moral sphere, unlike non-living entities.[103] Where the requirements for the good life are similar, different organisms should be treated equally as a principle of justice, but where the requirements are different, they should be treated differently. Verhoog evaluates genetic engineering in terms of what it means for the quality of our relationships with the modified organism. This includes damage to its integrity and wellbeing, the indirect effect on other relationships, and the loss of the appreciation of the individual. This last, he argues, is caused by the reductionist treatment of animals as objects in a process, which is necessary in order to do genetic engineering. Verhoog concludes that 'no genetic manipulation of animals should be carried out, unless a basic or very serious human or animal interest (question of life or death) is involved which cannot be met by other means.' As with Fox and Rolston, it could be questioned why so strong a line is drawn at genetic engineering and not other applications of science to animals, but there is force to some of Verhoog's arguments.

Andrew Holland[104] criticises the view that genetic engineering tends to view animals as instruments, by adopting Kant's assertion that we should not use other life forms 'as means only'. It is possible to treat another life form as a means, as long as it is not only as a means. Hence it may be possible to use an animal as an instrument, and yet at the same time also give it respect. Holland also argues that the potential to harm animals is broader than just physical pain, for instance if an animal would be rendered incapable of reproducing. The genetic engineering of animals should therefore not result in reducing an animal's capacity to exercise options. It is important to note that this same logic would apply equally to some aspects of selective animal breeding.

An Alternative Perspective

This brief survey reveals a wide variety of views of what human beings may do in relation to animals, based on different conceptions of animals, humans and the relationship between them. Problems can be seen in each case with the use of rights, utilitarian or ecological perspectives. Christian ethics attempts to understand the moral purposes of God for creation through the 'book of nature', the scriptures and the incarnation of Christ. Being cruel to animals is wrong, not only because these acts are ugly or because of their consequences, but because they are opposed to God's mind and will. Northcott argues that the Old Testament offers a fundamentally interactive account of the relationships among humans, the social order and the natural ecological order, and between all of these and the being of God.[105] This relational understanding offers a significant contrast with modern ethical individualism and subjectivism, and also with some of the perspectives discussed above. The Hebrews believed that moral values and purposes were enshrined in the nature of created order. Similarly the Christian doctrine of natural law represents a belief in the moral purposiveness and relationality of the cosmos, and in the relation between the human quest for the common good and the good of the created order and other creatures.

On this view, the moral life may not be reduced to individual human intuition and emotions, nor may moral judgements be limited to human experience and society. Rather the physical reality of created order, the community of human and non-human species and the ends and purposes which they differently serve, are given in the nature of the creation. Instead of the language of rights, the treatment of animals is set in a context in which they share the divine inspired character of all life. It expresses an understanding of the human responsibilities and duties towards animals that therefore arise, as Barclay also argues.[106] This is illustrated by the fact that the principle of life was seen to be shared between humans and animals, and was believed to reside in the blood. This was why the Hebrews were told not to eat the blood of animals they killed.

In this context, the Old Testament laid down moral duties with respect to both domestic and wild animals. The first humans are represented as vegetarians, but following human rebellion, the covenant with Noah shows God allowing a situation in which animals fear humans, since humans may thereafter kill and eat them.[107] From the context, this could be interpreted as less than an ideal situation. The Old Testament reflects a settled agricultural society in which animals have been domesticated. Domestic animals are to be treated with respect and compassion. Cruelty to them is condemned. They are not to be overworked, they are to be properly fed, and even the animals of an enemy should be treated kindly. In ritual sacrifice, the conditions laid down by the law for animal slaughter also reflect a measure of compassion and respect. For example, mother and young are not to be killed in sight of each other.

Other examples could be cited, but a fundamental idea is reflected in these beliefs and ordinances concerning the value of animals – even animals which would be killed and eaten. The taking of life, including animal life, should only be done for serious reasons and in a manner which reflects the moral significance of the animal's life in the prior context of God's created order. It is not *carte blanche* for any and every use, but it does allow for many valid uses, including the killing of animals for food. How then is this to be expressed when it comes to human benefits from modern technological ways of using animals, including genetic engineering?

Human versus Animal Benefit

It is helpful to pose the question the other way round and ask what is the basis for the notion of human benefit, and what are its limits. Under virtually all ethical systems, except possibly those of the Jains, human life is generally reckoned to be of greater moral value than non-human life. In the Hebrew and Christian traditions the supreme moral value of human life is underwritten by the doctrinal belief that humans are more God-like in nature than other life forms. This is expressed in the idea that humans reflect the 'image of God'.[108] As discussed in Chapter 3, this has no single agreed interpretation. In the present context of human comparison with animals, the godlikeness of human beings has been variously identified with the human capacity for language and communication, for abstract reasoning, for complex and caring relationships, and for creativity, not least those creative powers exercised by the scientist. When we as human beings remake the world after our needs and desires we may be said to share in the creative action of God Himself who made the world.

If human life were seen as the supreme ethical good in the world, then the biological, social and intellectual enhancement of human life becomes the central task of any civilisation with a sense of ethical purpose. John Finnis argues that the pursuit of life as the central ethical good involves a quest for all those things which are essential to a good life including adequate nutrition, freedom from pain, sexuality and the procreation and nurture of children. They in turn experience life as a gift from their parents, and ultimately from God the creator.[109] As we have seen, however, such a strong emphasis on human good must be modified in the light of the wider relationships God has set in creation. The three transgenic animal case studies considered in this chapter provide a useful framework in which to examine this question, in that medical benefit is the primary goal in each case, but with an increasing scale of intervention. There is a gradation from animals as producers of medically useful chemicals, through animals as sources of spare part organs, to animals engineered genetically to exhibit human disease in order to act as scientific models and as test beds for potential therapies.

Two Sets of Assumptions and Two Cultures

The debate has so far been framed in philosophical or theological terms, but before examining it in more detail, it is important to recognise that there is a social element also. The issues may be stated from the viewpoint of two social groupings, namely the medical community or those concerned for animals. These reflect, in a sense, two entirely different cultures, each approaching the problem from its own set of assumptions. One is primarily concerned with experiments to relieve human suffering, looking for the best ways of doing so, and perhaps trying not to think too much about the effects on animals, or at least rationalising them to a certain degree. The other starts from its concern over what is being done to the animal, and tends to make little of the consequent result for humans. The respective weight given to arguments concerning human suffering and animal suffering is likely to depend on which approach is seen as primary. This may in part arise from which situation people are most exposed to. People who are continually faced with human suffering are likely to be more tolerant of what is done to animals in order to alleviate this suffering. Those whose view is more framed by an awareness of animal suffering at human hands are likely to be far more critical of the medical benefit. Obviously, it would be misleading to suggest a complete, black and white distinction, but it highlights an important point about emphasis and the social reinforcement of that emphasis which occurs within each culture.

This review of underlying perspectives on animal–human relationships has already demonstrated the complexity of this key issue, which this chapter now seeks to explore in more detail.

ANIMAL USE AND MISUSE

Animal Suffering

The previous chapter identified a mixture of animal welfare implications in the four relevant case studies, which are now summarised. To produce pharmaceutical proteins in milk of animals seems in many, but not all, cases to be relatively innocuous in terms of harm done to the animals, certainly once the experimental stage is past. It is unlikely to fall into the category of a degree of harm which should never be inflicted. In the discussion of xenotransplantation, it was argued that there were no special reasons to consider that a pig's welfare would be compromised during its life, although the need for a disease free environment may result in barren and less desirable conditions for the pig. There would always be the temptation to cut corners on welfare, in either case, but the need for a pristine animal is sufficiently high, and the outcry if something went wrong is a sufficient incentive, for the best practice to be the norm. Serious intervention would be necessary to extract the pig's heart, but the animal would be anaesthetised and would never recover consciousness. In either case, if nuclear transfer techniques were used to enable more precise genetic modifications to be performed it should mean that less animals were used experimentally, but there may prove to be some welfare problems associated with the technique, as mentioned in the previous chapter.

The oncomouse case is of a very different order, in terms of suffering. If no impairment of the animal's metabolism occurred because of the cancer then its value as a model for human disease would be questionable. It is the similarity between the animal model and the human situation which forms the basis of its value as a model. There is variation in the degree to which humans who contract cancer experience pain and the other forms of distress associated with cancers, and it is of course difficult to gauge exactly what an animal is suffering in each particular instance. Inevitably, our perception of what is meant by animal suffering is to some extent an extrapolation from our human experience. The point goes beyond academic debate. There are clear signs that animals can and do show distress and feel pain, and since we cannot communicate in order to find out, it is only fair to give them the balance of the doubt.

The God-given mandate of care towards our fellow creatures would point to a need to be precautionary over our uncertainty in this area, not *laissez faire*. It would seem reasonable to suppose that the genetic modification of mice to exhibit serious disease characteristics is likely to cause significant suffering to a large number of mice.

The much debated question of animal sentience and suffering is not the whole case. The fundamental point of whether it may be considered wrong to induce a disease in the animal, regardless of the animal's level of sentience, is discussed in the next section. This goes beyond the more familiar concern over the degree of suffering which might be involved, and poses a more metaphysical question. Each of these cases is an entirely novel use for the animal. Do any of these cases represent an appropriate way of using an animal in principle, irrespective of the matter of suffering as such?

Right and Wrong Use of Animals – Pharmaceutical Proteins in Milk

A strong positive ethical case could be made for the production of pharmaceuticals in the milk of animals, Case Study 8. An animal is being put to a potentially extremely beneficial human use, with a relatively small intervention which does not compromise the animal's integrity. On the other hand, it could be argued that to engineer sheep and other mammals as living bioreactors for pharmaceutical production is an instrumental way of using the mammary function of the animal. Is this unacceptable? Humans have used animal milk as food since time immemorial, and intervened to a considerable degree in the animal's pattern of lactation and suckling its young. We already use the chemical production facility of their mammary glands to produce milk and milk products, like cheese and yoghurt, for human consumption. These adaptations of a natural function of the animal are not normally seen as ethically unacceptable. For some it is more acceptable than killing an animal for food.

On this basis, it would be hard to argue that to produce a particular protein in milk represents an inordinate change of use, especially in the case of Tracy: what she produces is the human version of a protein whose sheep form she manufactures anyway. Admittedly the sheep version of the gene expresses the protein in the liver, and thence into the blood, whereas the human gene expresses in the mammary gland. The main objection would be if it was deemed unacceptable to change the animal's metabolism for any reason at all, or at least one that was not in the animal's interest. More concern might be raised if a protein were expressed in an organ or at a level which resulted in damage to the animal, as in the erythropoietin case, or a protein toxic to the animal, but these might also be ruled out on welfare grounds.

Right and Wrong Use of Animals – Xenotransplantation

Intervening to modify and then take an animal's heart is a much larger step. Again, this needs to be seen against the wider context of our relationship with animals. Xenotransplantation unquestionably represents a radical extension of this relationship. The use of animals as donors of organs is fundamentally different from well-established uses discussed above. If it were now seen as acceptable to start producing animals to supply spare parts for the human body, then some would argue that we would be succumbing to an unduly instrumental view of the animal, one which goes considerably further than making pharmaceuticals in its milk.

That this is not an idle concern is revealed by the terminology which is used. In calling an animal a bioreactor, or a spare part supplier, it is described primarily for what it is functionally – as a means to an end – not what it is as an animal. The functional analogy with chemical production and mechanical engineering gives a clue to an underlying trend, a change in the way in which we look at animals. By extrapolating from concepts of the factory, a statement is made that they are more closely related to the non-living world than the living. This phenomenon is discussed further in Chapter 8 in the context of patenting living organisms.

Some argue that the techniques involved in heart xenotransplantation are just an extension of existing procedures like the use of artificial heart valves from pigs, insulin from pigs, and skin from pigs. There are however some important differences. Valves and skin are specially treated to kill the tissue in order to prevent rejection by the recipient's body. The valve, for example, is used purely for its physical properties, as a piece of elastic, not as a living organ. Transplanting pig hearts would be using entire, living organs. A step change is involved. Moreover, even after transplantation into the human body the heart remains genetically a pig's heart. It does not become human.

The argument is a good example of gradualism. A way is discovered of using a mechanical device in the body, like a pin in a fractured bone, which does not raise undue ethical concerns. A small step is then made – using a pig valve. This is of greater ethical import, and the public may or may not have realised it has happened, but it is now quite a common practice. Transplanting an entire heart is then presented as being only another small step further – more of the same. What has happened is that it is being maintained that accepting the idea of pig valves has established the general principle of using any animal organs in people, even though the question of entire hearts may not have been in anyone's mind at the time. What amounts to a significant step change is made not to appear so. It requires us to step back and look at the complete sequence of steps and ask if the final end is in fact acceptable.

Perhaps the most common case put forward to justify this use of a pig's heart is the ham sandwich argument. If we rear pigs to slaughter them for food, how can we object to rearing them to save human life? Although the difficulty was pointed out in Chapter 3 of using the concept of 'natural' to evaluate an ethical position, there is nonetheless a 'naturalness' argument to be considered here. It is a basic fact of life that everyone has to eat to live. We may debate whether it is acceptable for humans to eat animals, but it is clearly natural in the sense that many animals are also carnivores. One cannot make the same sort of argument that no one should die because of not having a heart replacement. The possibility of a heart replacement has only come about through human skill, building on the strengths of centuries of medical, scientific and technological advances and understanding. It is not natural to use an animal as a spare part in this way. It is human artifice. That is not to say it is wrong, but in ethical terms it is not the same as eating an animal.

It is somewhat surprising therefore to find that the Nuffield Council on Bioethics uses the argument that we eat pigs but not primates as the main ethical justification for its conclusion that it is acceptable to use a pig's heart for xenotransplantation but not a primate's.[110] This seems rather a weak basis on which to build such a fundamental distinction amongst higher animals. Leaving aside for the moment more pragmatic questions of safety, it may be asked what ontological basis there is for making such a distinction. If the decisive factor was that the animal is not a primate – then potentially any non-primate would do. Alternatively, is the crucial point the need for a certain measure of dissimilarity, in which case it could be asked if pigs are still too close to humans for comfort? To make an explicit distinction between primates and other mammals would imply a gradation in animals, and thus in what we may do to them. This then raises a question of what warrant there is for such a graded valuation of animals. In the gospels, Christ makes the observation that not a sparrow falls to the ground without God's will, yet human beings are 'worth more than many sparrows'.[111] One might infer from this common sense remark that there are also some animals which are more valuable than others, but not where a line might be drawn. Pigs are amongst the most intelligent and sociable of farm animals, and held in special affection by many people. Although primates have a much closer biological kinship to humans, it is by no means obvious that pigs are so much more remote as to justify removing their hearts for human spare parts. Indeed, this points to an awkward dilemma. If the animal is not reasonably closely related, its heart would presumably be of no medical use, but if it is too close then it raises intrinsic ethical objections. At best then, the pig is a compromise, and one which will not satisfy some.

Against this backdrop, a biblical understanding includes a duty of care for animals, as our fellow living creatures, since we are not only stewards of nature, but also its companions. There is a horizontal relationship which, though far short of human-human relationships, still has the vital dimension of dignity and respect for the creature. This precludes a purely instrumental attitude towards them. Intervening to take an animal's heart is a major step. Rather than being seen simply as a technological opportunity, the relational perspective suggests that it would belong to the class of things which we might not do to an animal, were it not for there being a serious human need in view. Moreover, it is one we should not embark on without having looked for all the alternatives, and had a reasonable expectation that it really would deliver a significant human benefit.

Right and Wrong Use of Animals – Transgenic Mice as Human Disease Models

If modifying an animal to remove one of its organs for transplant presents serious questions about inappropriate use, intervening to cause it to have cancer is still more problematical. The problem is closely tied to the suffering which could be involved. It might be asked whether there would be an ethical distinction amongst the many types of cancer and other forms of serious disease. For example, would it be ethically unacceptable to induce a fatal cancer in a mouse, but acceptable to induce a form of muscular dystrophy which is not fatal to the mouse? Behind the minutiae of levels of suffering, however, lies a more profound question.

This is the issue of principle which declares that it is simply wrong to create an animal which has no option but to suffer in such a way, no matter what the end. This is similar to the concept set out in the Banner Committee Report (Box 5.1), in assessing the ethical approaches to be made to emerging animal breeding technologies.[112] Before assessing whether or not a particular animal harm is justified by comparison with a human benefit, it has first to be asked whether the intervention falls into the category of 'harms of a certain degree which ought under no circumstances to be inflicted on an animal'. The example of unacceptable harm given was the hypothetical modification of a pig to reduce its sentience and responsive-

---

BOX 5.1 THREE PRINCIPLES FOR THE HUMAN USE OF ANIMALS IN THE BANNER REPORT[112]

1   Harms of a certain degree and kind ought under no circumstances to be inflicted on an animal.
2   Any harm to an animal, even if not absolutely impossible, requires justification and must be outweighed by the good which is realistically sought in so treating it.
3   Any harm which is justified by the second principle ought, however, to be minimised as far as is reasonably possible.

---

ness. It was argued that even if there were no welfare implications to this modification, it would be morally objectionable because it treated animals as 'raw materials upon which our ends and purposes can be imposed regardless of the ends and purposes which are natural to them.'

Does inducing cancer genetically in a mouse fall into this category? Some maintain that the genetic component is unimportant. If the intention is to induce cancer in a mouse anyway, it is argued that it makes no difference whether this is induced by chemicals, irradiation, genetics or by other means. Indeed, there is already an underlying ethical issue of whether it is right to induce cancer in a mouse by any means, which raises questions about the acceptability of existing practices.

Nonetheless, the genetic route does appear to make a difference. The very advantages which are claimed for oncomice over non-genetic inducement – the reliability and repeatability of the effect – make it the more certain that the mouse will suffer adverse effects. To the extent that genetically engineering the mouse is the most effective way of inducing a tumour, where administering a chemical carcinogen or even selective breeding would leave a greater uncertainty, then it becomes more of a problem ethically. To the degree it can be made sure that a given mouse will indeed develop a tumour, the more there is an ethical case to answer, for, in effect, forcing that tumour on that mouse. For some, indeed, the genetic route is more repugnant because it achieves its end by making the change at the most fundamental level of the animal's genetic make-up, and is therefore predisposed from the moment the animal is conceived. This is seen as different from those methods in which a given mouse would not develop the cancer unless the carcinogen was actually administered. In a laboratory situation, many reasons could be envisaged why the administration might not happen in practice – such as equipment failure, loss of vital funding, or a decision not to go ahead with a particular experiment. In such a situation, that individual mouse would not develop the cancer. If, on the other hand, the mouse had been the product of a germline genetic modification, or homozygous breeding, it is internally programmed so that it has no chance but to develop the cancer.

As already discussed, the Old Testament sustains a strong ethic of care towards animals, which limits our use of them, to exclude mere greed or abusive purposes. From this it is a natural extension that gratuitous or excessive suffering would be outlawed. On this basis, from a Christian perspective there would be a strong *a priori* argument against the right of humans deliberately to induce cancer in another organism, no matter how morally compelling are the reasons for doing it.

Right and Wrong Use of Animals – Cloning

The achievement of Wilmut and his colleagues[113] in cloning a large mammal from the cells of a mammary gland represented a major shift in the accepted understanding of animal cell differentiation. Animal cloning had been possible to a limited degree by splitting embryos, itself not without ethical concerns, and had been practised only to a very limited extent, mostly in cattle. At the time of writing, the new method has now succeeded in various cell types in two breeds of sheep, in cattle foetal and oviduct cells and recently also in mice somatic cells. Rather little is understood of how it has happened, however, or which cell types or animals it will work in. Different farm animal species can differ quite markedly in reproductive interventions like artificial insemination and embryo transfer. It remains to be seen how generally it can be applied to different farm animals without adverse effects. Assuming that it can be applied more widely, the Roslin work has now opened up the prospect of a range of applications of cloning by nuclear transfer, as discussed in Case Study 11.

Important ethical questions are raised about the appropriate use of animals. Firstly, there is an intrinsic objection that it is contrary to something fundamental about life. One of the most characteristic features of the created order is its variety. Throughout the Bible, in commandments, stories and poetry, the general picture is of a wonderfully diverse creation, whose sheer diversity is a cause of praise to its creator. It could be argued that to produce replica animals on demand would be to go against something basic and God-given about the nature of life. It reduces the living organism to a narrow blueprint, when in general mammals rely on maintaining a certain level of genetic diversity through sexual reproduction. Indeed, the very fact that there are genetic limits on how far cloning would be useful to animal breeders reflects something about the nature of things. The presumption that human beings know what is the optimum, on the basis of a very narrow range of traits out of the vast diversity and complexity of the animal, is seen as an act of hubris and irresponsibility. Taken to the limit, this would mean that cloning an animal would be inherently wrong, no matter what it was being used for. This is an intuition which runs deep in many people.

There is, however, a question of scale and intention to be taken into account. Cloning might thus be acceptable in certain very limited contexts, such as research, where the main intention was not the clone as such but to grow an animal of a known genetic composition, where natural methods would not work. The primary aim of the nuclear transfer is to find a much more precise way to do the type of genetic modification represented by Tracy the sheep in Case Study 8. The fact that the method produces clones is something of a side issue. This factor would, however, be exploited to produce perhaps five or ten animals from a single genetically modified cell, but these would be bred naturally thereafter, since there would apparently be no advantage in cloning beyond this first point. In this context, the limited use of cloning to produce a few founders of a set of lines of genetically modified animals for small scale medical applications would not seem excessive, a view accepted by the Church of Scotland General Assembly in 1997.[114]

The extension of cloning to somatic cells in both cattle and mice makes it clear that the range of specialist applications will broaden. In some cases, once a transgenic animal has been achieved, there will be pressure to use cloning as a deliberate means to replicate the animal. This might be more difficult to justify ethically. There will be a wide spectrum of research applications arising out of the achievement of cloning in mice, since mice are easier to use and better understood in a medical research context. At this early stage, one can only speculate on the particular uses. Some suggestions, like basic research into cell differentiation, would seem less problematical than applications to growing cloned human cells or even organs, which would raise major ethical questions beyond the scope of this book. It seems clear, however, that more mice will be used in research as a result.

What would raise more ethical problems would be the use of cloning in routine animal production, in order to side-step normal breeding methods on the grounds of economics or convenience. Those who do not see a fundamental objection to cloning argue that this is no more than an extension of selective breeding, and of methods like artificial insemination and embryo transfer. If there was a clear benefit to the farmer to start off with prime stock, in order to produce the best beef or pork, this might seem to have its attractions. As was observed in Chapter

3, this line of argument begs the question of whether what we are already doing is ethically acceptable, and Chapter 10 asks who actually gains the supposed economic advantages. A more profound challenge to this view comes in the wider context of how humans use animals.

Cloning is a singularly instrumental act towards an animal. To take an individual and seek to make as near an identical copy of it as possible could be said to be an abuse of the uniqueness of the original animal – a utilitarian perspective taken to the extreme. Serious questions would therefore need to be asked of what was driving the technology, and whether it is really necessary. How important is it that more animals reach more rapidly a particular level of what the current received wisdom regards as genetic improvement? A good ethical case might be made for developing disease resistance in the animal. Another case could be to prevent the extinction of an endangered species, if the critical factor for immediate survival was a simple question of numbers, which cloning could address, rather than one of genetic diversity, which it could not. For many breeding traits related primarily to perceptions of consumer preference, however, it is much harder to justify the intervention represented by cloning.

If the reason for cloning meat-producing animals is ultimately rooted in the demands of the supermarket production system, one has to ask if the end justifies the means. It raises the question of at what point in intervention we have to set a limit to avoid merely using animals in the same way that we would use microchips on a production line. Taken to the extreme, the economic ideal might be to manipulate animals to be born, grow and reach maturity for sale and slaughter at a time to suit production schedules, to provide consistent joints of meat of reliable quality on the shelves. The Church of Scotland has viewed such cloning as one step too far in putting the mechanical paradigm of mass production efficiency over allowing an animal the freedom to be itself.[115] This is a point at which humanity should hold back over what it could do technically, in recognition that we are called to show respect for the integrity of these fellow creatures, as more than just supermarket commodities.

A further development has been a line of research conducted in the US, in which nuclei for cells taken from various animals, including sheep, pigs and monkeys have been introduced into denucleated cattle cells.[116] This is only an embryological experiment, and it is scarcely conceivable that such a procedure could ever produce viable offspring, but it raises very serious ethical objections over the mixing of species in a gross way, rather than simply changing one or two genes. It has to be asked whether this is taking scientific curiosity too far.

HUMAN BENEFITS

The Case for Human Benefits

For the production of pharmaceuticals in milk of farm animals, as far as can presently be seen, the level of animal suffering and intervention is at a rather low level. The oncomouse stands at the other extreme, where the intervention done to the animal is so serious that a strong case *prima facie* could be made that it is unacceptable for any reason. Xenotransplantation stands somewhere in between. On the other side stands the case in favour of such animal interventions for the human benefit. This must now be examined.

The primary moral argument for the oncomouse and xenotransplantation is that the human need overrides all other considerations. It is recognised that there is a clash of moral goods, but the suffering of large numbers of human beings and the medical vocation to combat human disease take precedence, and justify the use of animals as models or for spare parts. In the extreme this would allow any use of animals, but a more measured view would recognise the notion of limits. But it would ask the question the other way round. Are there certain factors which are of such great benefit as to justify the sort of intervention and suffering involved in the oncomouse, for example, but other factors which are not? To help unravel this, a detailed examination of the case for human benefit is called for, before returning to the question of how the conflicting claims could be handled.

Human Benefits of Xenotransplantation

An important factor in the justification case is that the animal intervention really has a good expectation of delivering a substantial human benefit. For it to be ethical for humans to make the intervention in animals which are represented by xenotransplantation or mouse models, it is not sufficient simply to appeal in a general way to the goal of medical benefit. The benefits must first be specified and examined critically, for a society to weigh these against the animal suffering and intervention.

The case for xenotransplantation is a potential solution to the serious shortage in donor organs, and especially in the two most likely areas of application, namely hearts and kidneys. Even if everyone carried donor cards and the maximum use were made of human transplants, on current trends there would still be a large shortfall. Moreover, one of the main sources of hearts is from road accident victims. It can be argued that good is being gained from a human tragedy which will always be happening. However, it would be far better to reduce the amount of accidents. In addition, there is a clear ethical imperative to reduce the amount of overall road usage because of its increasing role in environmental damage. Significant success at either would reduce the number of available organs. Ironically this situation is a product of the success of the technique of transplantation. It is no longer limited by technical feasibility but by availability. The medical benefit of xenotransplantation would be the extension of the range of people to whom organ transplants could be offered.

These benefits are largely potential, however. The report by the Nuffield Council on Bioethics[117] draws attention to the many current uncertainties of the technique. Even assuming the complement hyperacute rejection problem were substantially overcome, that would not be the end of the story. A number of further steps is required to overcome such issues as coagulation and T-cell rejection, each of which may call for additional genetic changes in the pig.[118] Would the animal organs indeed successfully perform all the necessary functions in a human body? What side effects might there be? The Nuffield report makes something of a leap of faith with the assertion that 'many of the biological obstacles ... will eventually be overcome'. More caution is surely needed. The report may be correct, but in the early days when attention is focused on the potential of a new technique, there is a natural tendency to overstress its good points, when as yet some of its drawbacks may not have emerged. Historical precedent points repeatedly to the fact that novel technologies almost invariably turn out to have unsuspected problems.

It is important, however, to have an appreciation, if all these barriers were indeed overcome, of *how* good transplantation would be. The Nuffield report's argument in favour of xenotransplantation begins by citing the 'increasing success' of transplantation. It is important to clarify, however, exactly what is meant by 'success'. It means one thing to a medical researcher who is much encouraged by the greatly increased survival time and life quality, compared with the early days of heart transplantation. But this is a very qualified sense of success, compared with the more common usage of the word – which might expect it to mean that most patients return to a normal and full life up to their normal life expectancy. This is a quite different sense from statistics which are presented in terms of a person having a 50 to 70 per cent chance of living five years longer. There is a substantial difference between a cure and a remission. The patient will continue to depend on immunosuppressants. Benefit perception needs to be taken as much into account as risk perception. If the degree of recovery means the return to a high quality of life for most people treated, the case is much stronger than if it turns out to be a more hit and miss affair, when it would be harder to argue it outweighed the ethical objections.

There are also practical problems of careful matching. At this point in time we can not know, for example, whether certain 'lines' of animal organs would work better for some groups of people than others. If all went well, heart availability would certainly be improved beyond all recognition, but even so, the Nuffield report's claim that 'xenotransplants could be offered as and when they were needed' seems exaggerated. Quite apart from whether the health care system could afford as many animal organs as were needed, delivery could be constrained by the relative geographical location of patients and pigs. This is not a major problem in Cambridge, where the technology is being developed, but it could be a significant factor in more remote areas like the Scottish Highlands and Islands.

Human Benefits of Transgenic Mice as Human Disease Models

The need to advance human medicine is said to justify in principle the use of animals as models. If we are to make significant inroads into severe and distressing conditions such as cancers, such models are the best way to understand such diseases and test therapeutic remedies. This ethical judgement is at least implicit whenever the use of oncomice and other disease model mice is simply taken as read, and is probably widespread.

One problem which cannot be avoided is that, as with all animal modelling, there are limits on how far one species can model another. The closeness of the modelling will vary considerably from experiment to experiment and, to some extent, as the science develops. For the experimenter, the primary argument is that, though undeniably flawed, a mouse model is better than having a complete experimental gap at the whole organism level and it makes possible some things that would either be impossible or take a very long time to do.

An ethical justification given for creating mice which are almost guaranteed to suffer in this way is that by being more sure that a given mouse will develop a tumour, the data will be more reliable. If the efficacy is as good as claimed, then there is some force in this argument, but more is sometimes claimed for the potential advantages of genetic engineering than is delivered in practice. Groups opposed to animal experimentation contest the validity of mouse models. For example, a submission to the European Patent Office from two groups opposed to the granting of the Harvard oncomouse patent[119] cited a series of articles where the efficacy of this sort of testing was challenged. It also cited a leading US cancer research institute which has abandoned the use of mice in its search for chemotherapies.

An assessment of the evidence for each individual experiment or type of model mouse lies beyond the scope of this study. It is sufficient to note that serious doubts have been raised in the literature, even if it is not possible to assess how representative these are of the whole picture. It is clear that a crucial part of the ethical case in favour of genetically engineered mouse models rests on their efficacy. At its best an ethical case for real human benefit could be argued. At the other end of the scale, if an experiment is unlikely to work very well or it is for a relatively trivial reason, as with the hairless mouse developed for baldness treatment, it becomes much harder to justify the suffering to the animal.

From the medical point of view, then, what human ends would and would not justify the creation of such animals, taking account of the animal suffering which would result? There is a need for criteria against which to draw such a distinction, which is made the more pressing by the enormous range of applications of transgenic mice which have opened up. For example, would justifiable human benefits include all cancers, or only fatal ones, or say those with a significant incidence in younger people? Should the use be restricted to conditions which could not be studied by other means, or only to those which had a reasonable prospect of treatment? To draw up firm rules would obviously be difficult, but with the rate at which this area is expanding, it seems clear that some better clarification is needed. The temptation to duck the ethical issue, and rely on limiting mouse use just by the pragmatics of resource allocation, should be avoided. This is just the sort of case which an ethical commission on biotechnology ought to be examining.

Implications on the Use of Mice in Experiments

A major utilitarian argument is that to obtain the same data a few mice genetically engineered as disease models would reduce the need for many more non-engineered mice. This presupposes that research using mice is essential or will happen anyway. Substantial doubts have been raised, however, that fewer mice will be used. The result could equally well be that more experiments may be done, using the same number of mice as before. Indeed, evidence shows that there has been an enormous increase in the number of applications of transgenic model mice.[120] Not enough reliable data exist to be definitive, but it seems likely that overall there has not been a decrease in the number of mice used. It might be argued that more information beneficial to humankind is resulting, but this presupposes that all the extra experiments were equally worth doing. If model mice were rare and expensive it would focus the question of whether the experiment was really necessary much more than if they were routinely available. The use of cloning in mice will further increase the number being used in research.

In practice, much depends on the prevailing culture and assumptions in the research unit. If the general attitude is that the mice are just there to be used, as though they were items in a catalogue and not each a living creature of its own, much uncritical use of mice could result. This would be prevented if the culture of the laboratory is to avoid the use of animals wherever possible, and to require researchers to show that this has been attempted. The work of any UK research laboratory involved in animal experiments is scrutinised by the Home Office with respect to animal welfare, but there can be variability in how this is interpreted and applied. Moreover a Home Office licence is focused on welfare and is not meant as a guarantee of good ethics. It may be that establishments need to incorporate a degree of self-limitation, refraining from procedures which, while technically legal according to the norms of the day, were nonetheless ethically questionable.

Widespread use is possible partly because mice tend to be less highly regarded than larger mammals. Hitherto, the kind of genetic changes made in disease model mice using stem cell techniques have been proved impossible in larger mammals. The nuclear transfer methods which produced the cloned, genetically engineered sheep Polly have, however, now opened up a potential way to overcome this restriction. Attempts are being made to genetically modify a sheep to induce cystic fibrosis, on the basis that a cystic fibrosis sheep might, in this case, provide a better representation for the human disease than mice models. From an ethical point of view, such a development represents more than a simple step in a technology, especially given that modelling in mice can already pose a serious ethical dilemma. As with xenotransplantation, one would have to ascertain at what level in the animal kingdom it would be acceptable to perform genetic disease modelling of this nature, and how or indeed whether we should grade animals in terms of appropriate interventions.

Alternative Medical Strategies

An important question to ask where serious animal intervention is involved is to what extent there could be alternative ways of achieving the same ends. It should be asked how seriously people have looked for alternatives. Pragmatically, if model mice have been the easiest option, and there has been a research culture willing to accept the principle of using them, there has not then been the incentive to look elsewhere. In the case of mouse models, how great are the research benefits compared with what might be achieved by other experimental methods? The EU has put significant funding into this area, but as yet there are not enough data available to make a judgement one way or the other. Opponents of animal experimentation point to methods such as cell cultures which can be very effective in some circumstances. It is argued by others, however, that this is not a universal panacea, and that in any case it is much better to go from cell culture experiments to humans via an intermediate animal model stage. While the use of mouse models continues to increase, there appears to be recognition in at least some quarters of the need to reduce the use of animals in research, enshrined in the 'Three Rs' approach mentioned in the previous chapter.

Is xenotransplantation the most appropriate response to the shortfall in transplant organs? Attempts are being made to develop artificial hearts, although these appear to be at a very early stage. There is no more certainty of a successful outcome than with xenotransplantation. There might be alternative technologies to address the problem of heart disease, for example, to grow new heart valves and dermal skin. For some medical conditions, like diabetes, leukaemia or Parkinson's disease, one might grow modified cells or other tissues for transplant from cell cultures, perhaps using nuclear transfer methods. This would probably not apply to hearts, however.

Some also ask if it would not be better to spend the money on preventative medicine, pointing to the link between some cases of heart failure and an unwise diet and lifestyle. By no means does this apply to all cases, but it does raise a pertinent ethical point. If, to treat the results of what could be called human folly, a significant intervention in animals is being called for, there is clearly a duty to do rather more to prevent the situation arising in the first place by adopting more sensible ways of living.

There is also an 'opportunity cost' for this work. Standing back from the individual need, a utilitarian argument is sometimes put regarding novel technologies that the money could be used better in other ways to benefit more people. Are there better medical alternatives for the same investment of people, intelligence, innovation, time and money? Against this, the financial outlay in the research phase appears not to be very large, but if only wealthy people are likely to benefit from it, then there is a question of justice to set against the human benefits.

At the present time it is difficult to see what realistic alternatives there would be for the specific human need that xenotransplantation is aimed to address, but there is a moral obligation to continue to look for alternatives. There is also a need to consider how far one should go in pursuing any one particular medical benefit.

Human Risk

The production of pharmaceuticals in farm animal milk seems unlikely to raise serious risk problems, given that they must pass the normal clinical trials procedure to be licensed as medicines. The use of either the milk or the transgenic sheep itself as food is currently forbidden. The risk assessment of accidental entry into the food chain lies beyond the scope of this study but the ethical questions which this remote possibility might pose are more acutely addressed by the case of xenotransplantation. Here many of the human body's natural barriers to infection are by-passed by the combination of the direct introduction of animal organs into the human body and the genetic modification to suppress the normal defence mechanism against matter from foreign species. Furthermore patients would remain under a regime of immunosuppressant drugs which would reduce the body's disease resistance. The main concern is that pathogens not normally found in humans could be transferred. In particular, virus infections are difficult to treat and viruses can recombine to create new variants. In the UK, where Bovine Spongiform Encephalopathy (BSE) casts a long shadow, the focus has been on experiments which found that a pig retrovirus could infect human cells in culture. This has raised serious doubts in some medical quarters.[121,122] Given that the patient already suffers a terminal condition, this in itself might not be an overriding concern, but once the xenotransplant recipient was infected, the fear is that the disease might pass into the general population. In the US, where primates are currently the favoured source of hearts, the context is set by HIV and the theory that this may have been transferred from a monkey.

In the UK, the Department of Health has decided[123] that insufficient is known about pig retroviruses for it to be safe to proceed with clinical trials for the time being. The main ethical issue is weighing a remote risk of introducing a serious new disease to the wider population against the benefit to the large number of heart and kidney patients who would otherwise die. In either case, this would only be relevant if xenotransplantation actually worked. While it is right for the government to be precautionary, the amount of risk of a population effect from very limited clinical trials is much smaller than that from general clinical application.

Attitudes to Human Disease

One of the problems with either xenotransplantation or mouse models for serious diseases is the danger of making the potential medical benefits into a moral absolute. The objection in principle which some people have towards animal usage of this kind is perhaps more familiar than the opposite effect, namely that of elevating the ethical imperative to prolong life and ease suffering to a status which so overrides all other ethical considerations that it is close to being a moral absolute itself. The human dimension to such questions naturally pushes us towards the relief of human suffering. At one level, there is no answer to the argument expressed by such a question as 'If it was your friend who was about to die in their thirties, leaving a wife and small children, would you think twice about killing one pig, to provide a heart that might save him, or giving a group of mice cancer to test a potential cancer remedy?' But the foregoing discussion shows that this actually begs many questions about efficacy, resources, justice and wider human risk, quite apart from the use of animals, which must also be taken into consideration.

Ironically, the relative success of transplantation means that its potential may never be matched by the resources necessary – whether organs, money or health care facilities. The frequently stated situation of the shortage of transplant organs is itself a value judgement. It assumes a moral imperative that we should not leave the potential of a treatment of illness unmet, given that we had a means to do so. This creates a pressure which can then marginalise other important ethical criteria.

Not all technically possible human uses of xenotransplantation would necessarily be ethically justified when weighed against the level of animal intervention involved. Examples of this might be where there is limited efficacy of the procedure or its use for non-terminal conditions, or in conditions where there was an alternative therapy available, or using organs from primates.

The prospect of xenotransplantation is thought by some to change our view of human life more than it affects animal welfare. Some would argue that we would be stepping over a mark into animal to human transplants, in presuming too much on human use of animals. According to this, we should we stick to human–human transplants, with all the restrictions that carries with it, as part of the natural limitations of our humanity and its lifespan. This view would not be shared by most of our group, but it does raise profound questions about how far we should go on trying to replace failing parts of the human body.

SOCIETAL ASPECTS

Attitudes towards disease, risk and our valuation of certain animals compared with others, all have a strong societal element to them, but there are some more general social factors which should also be borne in mind in weighing up these issues. These include our intuitions about what repels us and what is familiar, associations of certain parts of the body, and commercial factors.

Intuitive Repulsion – the 'Yuk Factor'

A frequent reaction on first hearing of xenotransplantation is a sense of repulsion, the so-called 'yuk factor'. Some dismiss this as mere emotion, inferior to rational argument, but this seems to miss two important points. People do not make moral and ethical judgements as if they were simply a matter of cold logic. Moral intuition may often play an important or even decisive part, especially in recent years. (See Chapter 10 for further discussion on what we mean by rational and irrational.) Moreover, there has to be something people feel 'yuk' about, even if it is hard to pin down. The sense of repulsion is one way our culture makes judgements, but this is clearly something that is socially conditioned, particular to a certain time, certain groups, certain countries, and so on. Biotechnology which may be acceptable, for example, in Europe may not be in Japan, or vice versa. Specific things focus our attention today which were not so important perhaps a generation ago – like our concept of the body and its boundaries and taboos, and our concerns about risk and hidden side-effects. The notion of introducing a pig's heart can thus raise connotations of polluting the body through breaking boundaries. It is also noteworthy that we eat animals but not humans, but we are more dubious about having an animal organ transplant than a human one.

In the opinion of most of our group, these perceptions do not in themselves constitute a decisive argument against xenotransplantation, but it helps provide a context of where some of the feelings of repulsion may come from.

The oncomouse can elicit a similar response, but it is perhaps a less immediate example. For this, as for xenotransplantation, much can depend on how the issue is presented, for example in the media, or by proponents or opponents of the technology. It can be presented in terms of a miracle of science that someone who has suffered a major trauma is now able to walk around because of this wonderful technological innovation. It might, on the other hand, be presented as yet another example of natural boundaries being abused by technology driven blindly, without sufficient care for the consequences to you and me, or to the rights and feelings of animals. The sense of suggestion, either way, is very powerful. One especial concern is the tendency for exaggeration – either to raise expectations to vulnerable sectors of society of cures which might not be fulfilled, or the raising of alarms to the public at large which are only partially valid. Claims that xenotransplantation was ready for testing in humans once there were signs that the complement problem could be overcome have now been challenged as premature.[124]

Familiarity

Some argue that concern about xenotransplantation is primarily a case of unfamiliarity. On thinking through the issues, people would surely see the benefits of a mouse model or conclude that there is little difference between killing an animal specifically to obtain one of its organs and killing it to eat it. The problem is that we are simply less familiar with the new ideas. With many technological developments, unfamiliarity has indeed been a barrier to the society to whom it was first announced. The next generation has often grown up better informed and used to the idea, and often tends to look back on earlier attitudes with surprise or even condescension. Heart transplants are themselves an example of this effect. The sense of 'yuk' has in this case been largely dispelled by familiarity. But this is not the full story.

This argument is only valid in retrospect, as regards any particular piece of technology. To assert confidently that we will become familiar with it, and so accept it, is merely a prediction or a faith statement. In reality, there is no guarantee at all that will happen. The case of nuclear power, for example, has tended to go the other way. Indeed, there is an element of manipulation in the argument, since it requires the acceptance of the person that it will become familiar in order for it to become familiar. To seek to persuade someone that their present concerns will look silly in retrospect is socially dangerous. The more fundamental the impact of a new piece of biotechnology, the less certain it is that public acceptance is only a matter of time and familiarity, education and persuasion. Education and familiarity are not always the answer. It may be that a more fundamental intuition is at stake which people might never be happy with. A recent widely-publicised experiment in which a mouse had a human ear grown on its back seems to be such a case, and it is not impossible that xenotransplantation may also turn out to be unacceptable for some people, no matter how familiar they are with the idea.

Other Societal Perceptions

Issues of public perception in biotechnology are important, should be treated with respect and not simply dismissed because they may contain elements of ignorance, emotion and cultural conditioning. We are all members of the public. In many cultures, the heart has a special resonance. In the English language, drawing on a background of the Old Testament, it is the seat of emotions (especially love), courage, enthusiasm and innermost thoughts. It might be argued that the heart is felt to be too vital to what it is to be human to allow substitution by an animal's heart, but this would not seem to be an overwhelming objection. Many Christian and other perspectives would nowadays stress the concept of the body as a whole, and so would not relate our essential humanness to any one part of the body. One would require a very good reason to have a non-human heart, but it would not *a priori* violate one's humanness.

To put the xenotransplantation case in the extreme, one might ask the theoretical question of how much of a person would have to be made up of animal organs, blood and tissues before that person would cease to be human. Pragmatically, one might answer that long before a stage had been reached where this question had much reality, the life would probably be non-viable. But on the underlying point, from a Christian perspective (and in most traditional societies), the person is indivisibly mind and body, and does not cease to be that, notwithstanding a wide range of physiological changes, disabilities and injuries.

Commercial Factors

The commercial factor is one element of human benefit which is especially problematical and, whether rightly or wrongly, elicits much public suspicion. The notion of applying animal cloning to meat or milk production seems to be motivated primarily by production factors and consumer preferences rather than a more clearcut human good. For many people this would probably seem a less convincing reason than transplant organs or testing for cancer therapies, even though these too have their commercial side. In the case of xenotransplantation, there is considerable interest from the companies which supply immunosuppressant drugs. Some proponents have even suggested that the hearts might be provided free, because a company would make its money from the drugs, though this remains to be seen. The more technically innovative a new therapy, the more it is likely to require commercial participation. That is not a bad thing in itself, but unless it is made readily answerable to the public, it can lead to ethical priorities which owe more to the shareholder than to either the public good or animal good. This is discussed more fully in Chapters 8 and 10.

Conclusions

Balancing Ethical Goods

Exploration of the numerous aspects of the animal use and human benefit has identified a basic opposition of ethical goods – between what is appropriate use of animals and justifiable human benefit. From whichever side the issue is approached a near watertight case can be made, until it is appreciated what would then be the consequences for the other side. Clearly there is no simple answer. The existence of two separate cultures – of animal concern and of medical advance – should not be underestimated. For either side to claim absolute priority, and to insist that the other should give way, would seem unsustainable. In examining the detail of the case studies a number of points have emerged which indicate that the cases themselves are not as clear cut as might first be supposed. Each may well admit to concessions from their ideal position. To what extent do these show ways forward to resolving these two conflicting claims for ethical priority?

Unless the extreme position is adopted that rejects all use of animals by humans, distinctions need to be made over the degree of animal intervention which is acceptable. This involves some element of assessment of the consequences. In the case of the production of pharmaceutical proteins in the milk of animals, the intervention is relatively small and does not present a serious challenge to the animal's metabolism and well being. At the other extreme, however, a strong case has been made against the right of humans deliberately to induce cancer or other terminal illnesses in another organism, for use as a test bed for human disease, no matter how morally compelling are the reasons for doing it. It could be argued that this might fall into Banner's category of unacceptable harms. However, the consequential argument is that the animal intervention cannot be regarded in isolation from the fact that the only reason this is done is to give vital information about very serious human diseases. The human condition would not be made worse by ruling out this use of animals, but the prospect has been opened up of relieving human suffering of a type particularly poignant and distressing to our society at this point in its history. Having largely seen off epidemics of killer diseases which ravaged the population in past times, cancers and slow onset degenerative diseases have evoked the compassion of our times as being amongst the most pressing of human sufferings.

Faced with human death and suffering, the question of prior relationships is arguably the decisive factor. We regard the death of a human as far worse than the death of an animal, because humans mean so much more to us. There is an important difference here. We are more sensitive to our relationships with those individuals whom we might lose than to our relationship with the generality of animals. It is not a question of our relative versus our pet, but versus a remote batch of experimental animals that we will probably never know anything about. To disallow the prospect of any therapies for serious diseases that would have depended on such models would be a hard thing indeed, given our normal view of prior relationships. This is especially so, given that in the home we regard mice as vermin and are prepared to poison or trap them on the grounds of the risk to health.

Conclusions on Genetically Engineered Mouse Models

There are valid arguments to support both the medical case for using mice as models of human disease and the contrary case that animals should not be manipulated in such extreme ways. Consequently, it would seem that to take either of the extreme positions would require an unlikely weight of argument to show why the other position carries so much less ethical clout that it can be literally discounted. Neither the potential alleviation of human suffering nor the actual imposition of animal suffering seem large enough consequences to justify being regarded as absolutes which override other considerations. To do so would require something close to an intrinsic position – either a purely instrumental view of animals, or the absolute prohibition of all animal experimentation and genetic engineering. If neither position is taken, a compromise of some sort must be sought.

Compromises imply a balance. It would appear more likely that the balance is too far on the side of excessive use than of excessive caution. Arguably, the development of these model mice has proceeded too fast, without proper ethical forethought, and possibly without effective control mechanisms. It would be difficult to maintain that too little use of them has been made out of concern for the mice. It is not possible to be dogmatic, but the very rapidity of the development of model mice would suggest that the driving mentality has simply been whether it was technically feasible.

The problem is, however, that once the principle of transgenic mice was allowed, the flood gates opened to all kinds of transgenic mouse models and knockout mice for identifying genes and genetic function. It may be challenging an entire research philosophy, but there are serious concerns about what appears to be the routineness of such use of an animal, to the point where they hardly seem to have any worth. Even the attitude 'it's only mice' must surely have its limits, before they are treated as though they were not animals at all. There may be some logic in allowing more to be done to mice than to pigs or to primates, but this should not mean there are no limits for mice. It must therefore be asked how serious a disease and how much research justifies an animal suffering in this way.

We readily admit the need for the strictest limitations on medical research on live human subjects, notwithstanding the fact that such research would no doubt be extremely useful in understanding disease. There are thus limits to what may be done, notwithstanding there being a potential human benefit. One fundamental prerequisite of work on animals is that there is a reasonable hope of significantly reducing human suffering. This is already implicit in the Animals (Scientific Procedures) Act 1986, and was upheld in the Banner Committee report in its second ethical principle, that 'any harm to an animal ... requires justification and must be outweighed by the good which is realisitically sought in so treating it'.[125] It is also raised in the challenges which have been made to the value of mouse models.

If this perception is correct, some readjustment is needed. It needs first to be established that alternatives do not exist which would make transgenic model mice unnecessary, and then to limit their use to experiments for which there is no realistic alternative, even though the researcher has tried to find one. It is very clear that this will vary widely amongst mice, diseases and experiments. If model mice could only be used rarely (say, because of cost, or scarcity), other options would be sought, and the mice would be reserved only for the most essential experiments. There needs to be a culture of restraint to counter the temptation to regard model mice, whether knowingly or unthinkingly, as merely off-the-shelf items from the stores catalogue, of no more significance than chemical reagents. It would seem that there is a serious case to answer that as things stand at present the mice involved have been reduced to little more than material commodities.

Conclusions on Xenotransplantation

Xenotransplantation is a novel and serious intervention in animals. It does not raise such serious questions of suffering as the oncomouse, but it represents a change in the way we have related traditionally to animals. It would be unacceptable on principle if it was seen to represent a denial of the sanctity of both human and non-human life by mixing organs between species, or on consequential grounds if it was deemed an excessively instrumental use of the animal. The former would probably reflect the views of some Christians as well as others, including one of our group, but it is noted that in faith communities where the pig is regarded as ritually unclean, the use of xenotransplantation to save life may override normal prohibitions.

The consequential case is more complex. It is a debatable point whether xenotransplantation is taking away the animal's value in a way that killing for food does not, but a good case can be made that there is a significant difference between slaughtering an animal to eat in order to live, and slaughtering it for spare parts to prolong life. There are substantial objections in terms of animal intervention and also some concerns of loss of quality of life. The risk of transferring a serious disease to the general population is largely one of safety, but it raises an ethical point over what level of benefit to sufferers would justify what magnitude of risk to the population. Our relationship with animals suggests that xenotransplantation might belong to the class of things which we might not do to an animal, were it not that it might meet a serious human need.

In the light of these, the ethical case for xenotransplantation rests quite heavily on how effective it would actually be. If xenotransplantation is being presented as a well nigh essential use of animals, this should raise a query as to whether it will deliver the expectation. As with model mice, a prerequisite of a serious animal intervention is that there is a reasonable hope that the therapy will deliver a very significant reduction in human suffering. As discussed above, this is by no means proven. With such a novel technique, the benefits are still largely potential, and their true extent unknown. The primary motivation for the work is given as the long waiting lists for heart transplants, and that we have a moral obligation to do something about these. The Nuffield report, however, concluded that 'for the foreseeable future xenotransplantation will not solve the shortage of organs for transplantation.'[126]

At the present state of advances in this area, xenotransplantation is generally assumed to be a therapy only for patients in advanced stages of a life threatening condition, when no other treatment appears to offer comparable chances of survival. The massive invasiveness of a donor organ in the body, with all the attendant risks and also the constant lifelong need for preventative and supportive medication thereafter, is assumed to be a sufficient drawback for the procedure to be regarded as a last resort. In this context, in the case of the heart, the objections regarding animal intervention and other issues probably do not outweigh the potential benefit. This case would change if the technique proved to offer a substantial improvement of life and lifespan to most of the recipients, or there was genuine absence of viable alternatives, such as artificial hearts, which held out better prospects for success. Breakthroughs in the latter could tip the balance of acceptability, as could repeated failures of xenotransplantation. In such a case, there would eventually come a point beyond which repeated animal experimentation into xenotransplantation would not be justified.

Xenotransplantation should never be taken for granted, such that the animal became regarded – by the medical profession, healthcare administrators, those who reared the animals, or the public at large – in a primarily instrumental way. Consequently, if the use of pig hearts were acceptable, this should not necessarily be treated as a precedent which justifies all other applications of xenotransplantation. These would need to be considered on their merits, and bearing in mind whether xenotransplantation would be genuinely the best solution available. We should not simply assume *carte blanche* to take organs from animals in all other cases, given that we had once decided it was acceptable to do it in one case. Indeed, if it became routine there would be a danger in our very familiarity engendering a reductionist attitude, in which the whole procedure reduced the procedure of placing a modified pig's heart in a human body to putting a spare pump into a machine to replace a worn-out one. On such a basis anything could be exchanged, without any sense that animals were involved. Moreover, there would be a risk of xenotransplantation being used for more trivial reasons, non-terminal conditions or even non-medical ones. This is another case of gradualism. An antidote is to maintain a sense of acknowledgement of the intervention in an animal's life which is involved.

There are limits as to how far it would be justified to use an animal to preserve failing parts of the human body. Behind some of the discussion of transplantation there may lie an unconscious desire to pretend we can cheat death, that if we could only cure all diseases we could be immortal. The claims of some proponents of human cloning and the bizarre fascination in cryogenic preservation of bodies are manifestations of this. Part of the answer could be in whether we regard a heart transplant, human or animal, as a right or as a gift. If it is a right, this is claimed in competition with the rights of animals. A Christian understanding would see advanced medical therapy as an act of grace, something given which no one can presume upon. Moreover it sees life in the context beyond death, recognising that for all our remarkable medical breakthroughs death is not preventable, but that it is also not the end.

6 TRANSGENIC FOOD

INTRODUCTION

Two Major Issues

Because food plays such a central role in our lives, the genetic modification of foodstuffs raises more public concern and interest than almost any other aspect of genetic engineering. Anything which could affect the quality, safety or composition of our food matters greatly to us. The effect of the BSE crisis on attitudes towards eating beef in Europe was a vivid demonstration of how sensitive we can be to food safety, and also of how quickly public attitudes in this area can form, change and disperse. BSE and *E. coli* 0157 are non-genetic issues to do with food production and safety. In popular discussion, however, they have often wrongly been lumped in with concerns about genetic engineering. As a result, food technology issues, whether genetic or conventional, may have a considerable influence on the overall view of the population of the UK and Europe towards genetic modification as a whole. Indeed, where food is concerned, the more general questions about genetic engineering undoubtedly acquire a special flavour. Expressions like 'tampering with what is natural' relate to more than just debates about ethics and ideals, if we have to eat it. Now that genetically modified food has begun to be available in our supermarkets, with the prospect of many more products to follow, the debate about transgenic food is no longer an academic discussion of the possible issues of tomorrow. It is one of the realities of today which impacts on everyone.

Many see genetic engineering as an exciting and beneficial means to allow some current limitations to our existing crops to be overcome by the use of a much enlarged gene pool. Some see it as the only practical solution to feeding the world in the longer term. For others, however, it represents an intervention in nature which is either wrong in itself, or unacceptably risky when applied to food.

To be anxious about such issues could be said to be a product of the security of food supply which we enjoy in western Europe. Since having enough food is not in doubt, our focus tends to be on issues like safety and quality. Plainly, this is not the case for large parts of the world where merely to have enough food is often the overwhelmingly important issue, nor does it reflect the projected situation for the future human population as a whole. Where the context is survival, another group of questions relating to transgenic foods focuses on whether they have the potential to provide sufficient food for coming generations. Some are dubious that this is the right approach, but others hail it as the only realistic solution to an expected worldwide shortage of food.[127,128]

Food and Genes

In a literal sense, we are what we eat. Humans, like all other animals, are unable to synthesise directly the basic components of our bodies. As a consequence, we must obtain all of the basic materials that we require for our bodily growth and maintenance through the consumption of complex chemical materials, in other words food. The diet which we consume is conservative, in that it is derived from a relatively small proportion of the total number of organisms available on the earth. A high proportion of our food comes from the plant and vertebrate kingdoms, but we make use of relatively few species. As much as 93 per cent of the human diet is provided by only 29 basic crop species (eight cereals, three root crops, two sugar crops, seven grain legumes, seven oilseeds, plus bananas and coconuts), supplemented by about 15 major vegetable and 15 fruit crop species.[129] In Europe, most of our animal products come from just six species.

Normal digestive processes break down the complex molecules which make up our food, including the DNA, into relatively simple components. These are then resynthesized into complex molecules within our bodies. Because of this breakdown process, the genetic make-up of our food does not impact upon our own genetic make-up. The DNA contained within the food which we consume is not, as far as we are aware, directly integrated into our own genome, although concern has been expressed that DNA might, in some circumstances, be taken up by the bacteria of the human gut.[130] This seems highly improbable given that it does not seem to have occurred after millennia of eating non-engineered fresh produce.

Applications of Genetic Engineering to Food

For thousands of years, human beings have used the traditional genetic technique of selective breeding to modify and develop plant and animal foodstuffs. The range of genes which could be incorporated into a crop was restricted to those already present in the same or a closely related species. Transgenic methodology has opened up the possibility of going beyond the natural limits of the genetic separations which have evolved in the biosphere. In principle, genes can be introduced into a given crop or animal species from anywhere in the entire microbe-plant-animal-human gene pool. For example, we now have the techniques to introduce an Arctic fish gene into a cereal crop, which could enable it to flourish in unusually cold conditions, perhaps to the benefit of Scottish farmers. Although no one has so far done this particular genetic modification, one could imagine a guest at a Scottish hotel coming down to breakfast and being confronted with porridge made from oats modified with a fish gene, a case of 'Waiter! There's a fish in my porridge!'

Chapter 1 described the applications of genetic engineering which are being applied to agriculture and food production. These can be summarised as follows:

- to reduce the vulnerability of crops to environmental stresses;
- to enhance resistance to weeds, pests and disease, simultaneously using less chemicals;
- to reduce dependence on fertilisers and other agrichemicals;
- to increase the nutritional qualities of food derived from both crops and animals;
- to improve the taste, texture or appearance of food;
- to improve the yield from crops and animals.

There are also parallel developments aimed at producing novel substances, for non-food uses, in what would normally be food crops. Until the mid-1990s most of these potential applications were only at a research stage. Then in the UK in February 1996, the Safeway and Sainsbury supermarket chains began marketing a concentrated tomato purée which had been produced from genetically modified tomatoes. For most people in the UK this represented the real first impact of genetic engineering on foodstuffs, although genetically engineered rennet, using a calf stomach gene for the enzyme chymosin, had been used in the manufacture of cheese for some time. Other products, such as genetically modified baker's yeast, had been cleared for use, but had not been marketed because of public opposition.[131] Since Calgene's pioneering Flavr Savr™ tomato (Case Study 6) and Zeneca's paste, the picture has changed dramatically. Global acreage of transgenic crops increased almost fivefold in 1997 due largely to the production of herbicide tolerant soya and insect resistant maize in the US, and herbicide tolerant canola in Canada, see Figures 6.1 and 6.2. Companies from the US are now also seeking to put these into the European market. The resulting import of genetically modified soyabean and maize has, however, led to a heated controversy within the UK and the EU. This has focused not only on the issue of genetically modified food itself, but also on widespread concerns over the pressures which appear to be driving its introduction into Europe, against significant consumer pressure and the advice of some experts. It has also brought to a head important questions of food segregation and labelling, and the failure to allow a proper public debate on the issues, which are discussed later in this chapter.

The use of transgenic crops appears to offer considerable potential to enhance the security of the human food supply and to provide opportunities to reduce some of the adverse environmental impacts linked to

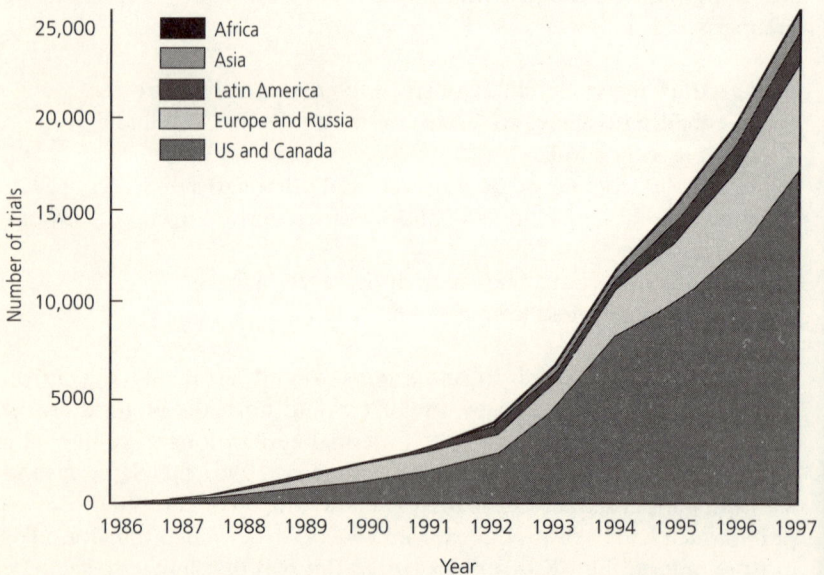

Source: James (1997),[132] reproduced by permission of International Service for the Acquisition of Agri-biotech Applications

Figure 6.1 *Number of Transgenic Crop Field Trials Worldwide, 1986–1997*

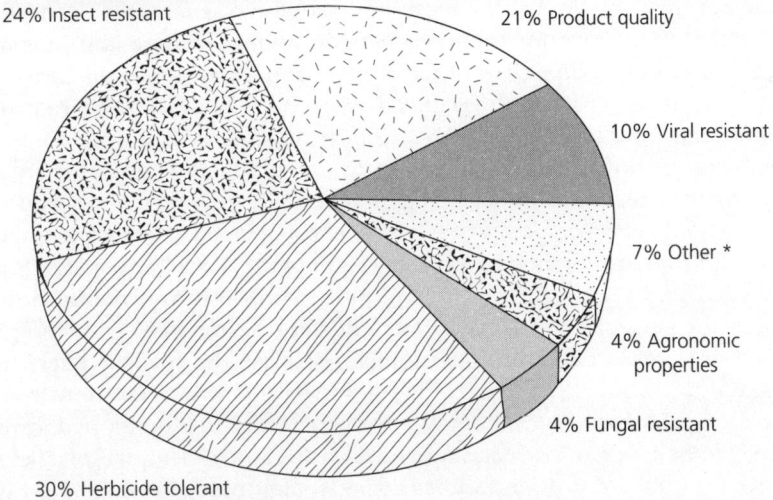

24% Insect resistant

21% Product quality

10% Viral resistant

7% Other *

4% Agronomic properties

4% Fungal resistant

30% Herbicide tolerant

* Other = marker genes, bacterial resistance, nematode resistance

Source: James (1997),[132] reproduced by permission of International Service for the Acquisition of Agri-biotech Applications

Figure 6.2 *Types of Transgenic Crop Field Trial Sites in the US, 1987–1997*

food production. There seem therefore to be powerful reasons for the rapid introduction of transgenic foods. The very power of these arguments, however, means that wider issues need to be assessed and be subject to informed public debate.

Ethical Questions Raised by Transgenic Foods

There are fundamental ethical issues about whether it is right to change foodstuffs in this way. For some groups in society, to embark on the sequence of steps represented in Figure 6.3 would pose a serious problem, which in turn raises important questions about how a democratic culture should respond. Transgenic food also poses wide and complex questions about the way in which our duty to provide food for present and future generations is to be achieved, and against what criteria the key decisions are to be made. In one school of thought, the question of feeding a rapidly expanding world population lays claim to being the supreme ethical consideration, before which all other considerations are of lesser importance. From a more environmental perspective, concern for our duty of care for the environment, animal welfare and human health and safety, on behalf of those same generations, weighs heavily. How are these two to be balanced when they conflict? The role of commercial and political imperatives in relation to public opinion and democratic choice and participation has emerged as a problem area, especially in regard to the labelling of genetically engineered foodstuffs.

It should also be asked to what extent crop genetic engineering is the most appropriate solution to food shortages in the developing world. If it does indeed have a significant role to play, then is this truly being borne out by the current priorities of investment and research in biotechnology? Does this amount to a hollow promise, because the focus is likely to remain the consumer markets of the industrialised world, far away from the areas of most need? Assuming techniques could be developed, care needs to be taken to consider how novel crops and agricultural practices impact on the cultures and values of the communities in which food is grown, their indigenous agricultural lore and their livelihoods.

ISSUES OF PRINCIPLE ABOUT TRANSGENIC FOODS

Underlying Questions

The basic question is whether it is inherently wrong to genetically engineer foodstuffs or their precursors. There are two aspects to this question. From the point of view of Christian or other ethical perspectives, is it right to modify food genetically or to use transgenic organisms as food? Does the method of genetic modification make any difference to such objections? (See Figure 6.3, step 2). Secondly, what is the moral significance of genetic material taken from one species, copied in the laboratory (see Figure 6.3, step 3), and used in another species as food? How does this affect such questions as cannibalism, if it was of human genetic origin; vegetarian concerns, if it was of animal origin; or religious strictures and traditions, if it came from ritually unclean species?

Given the religious and cultural significance of food, and a situation where particular faith communities or other groupings in society may be fundamentally opposed to consuming food of this type, there is an important corollary to these questions. How should a society cope with even a small minority holding inherent objections to a widespread practice? These questions were addressed by a Ministry of Agriculture, Fisheries and Food (MAFF) Committee under John Polkinghorne,[133] and a critique of some of its findings is given later in this chapter.

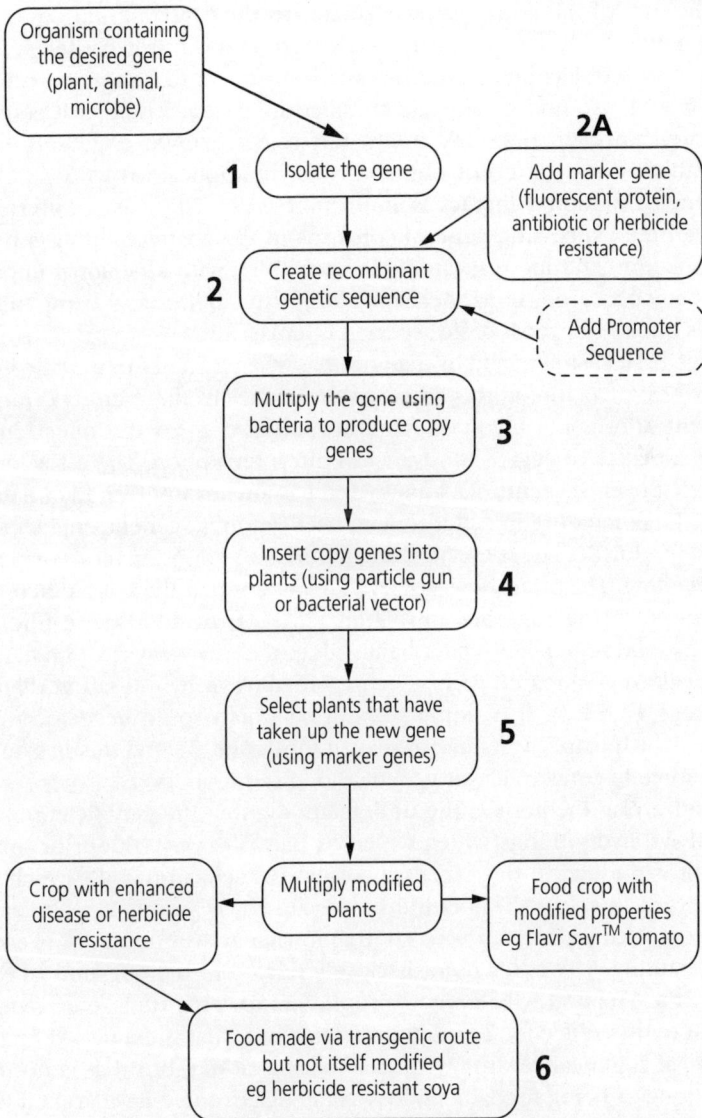

Figure 6.3 *Steps Involved in Plant Genetic Engineering*

The first of the above aspects relates to the fundamental concerns about genetic engineering itself which were laid out in Chapter 3. The force of such objections would automatically mean that those who hold them would also find genetically modified food unacceptable. It is quite possible, however, that one might not reject genetic engineering in general, yet have inherent objections to its application to food. The element of personal impact is important here. There is a difference between me expressing general concerns in the abstract about genetic manipulation, and me responding to something with a personal impact. As already observed, food, perhaps along with shelter and companionship, is one of our most basic aspects of human life.

One basic instinct, held by many, is that when it comes to food, 'nature knows best'. Examining this situation more closely, there are two rather different arguments involved. One argument regards the question as simply a matter of degree in respect of our intervention. The assertion is made that over the centuries humankind has already intervened in nature in very large measure in order to feed itself, and that genetic engineering is merely a further step beyond using natural mutations, manure or selective breeding. The other view, however, maintains that there is a difference in kind. To bring together unrelated parts of the total gene pool to produce food represents a paradigm shift in the way we work with nature in agriculture, compared with selective breeding or hybridisation (Figure 6.3, steps 1 and 2). It is doing something which could never occur in nature, which marks a significant reason for doubt. Behind this lies much of the force of concern about genetic modification in food.

Underlying themes are the understanding that the genetic composition of the living things which we eat is part of a wise blueprint set by God, or is implicit in the processes of natural selection, with which we should not interfere. This could be because it is simply not for us to change, or that we are not wise enough to change without harm of some kind resulting. The latter point is closely related to the question of risk. Part of the aversion which some people feel towards transgenic food is undoubtedly a reflection of safety concerns, but what is expressed here is its ethical significance. It may or may not be riskier, but that is not the same thing as being fundamentally wrong. Again, some have argued that these are not so much questions of ethics as of aesthetics, but this would be an unwise conclusion. Given the importance of personal and cultural taste where food is concerned, and the impact this has on our ethical perceptions, it would be unfair and even manipulative to dismiss ethical concern of this kind as mere personal taste.

Religious perspectives are especially important. Large sections of the early books of the Old Testament are devoted to the definition of allowable and non-allowable foods,[134] which include in some cases proscriptions against mixing certain kinds of crops and food animals,[135] on grounds of their ritual and symbolic significance to God's people. This clearly remains of profound importance to many Jews and there are related regulations within Islamic law. Christians, on the other hand, have generally regarded the ritual prohibition over the use of particular animals or means of food production as superseded by the life and death of Jesus Christ. Nonetheless, important indirect insights can be gleaned from the Mosaic laws on the general practice of agriculture, and the use, preparation and consumption of food, the relationship of human beings and their societies with the land, and questions of equity and justice. These relate to the ecological concerns and also some of the wider consequential ethical issues about transgenic food.

In contrast, it can be argued that a measure of freedom has been given to humanity by God, or is inherent in the human place in evolution. This could provide a justification for applying genetic engineering to food as to any other legitimate human activity. As human skills have developed it is entirely appropriate to apply them in this way, provided that other ethical and safety considerations are not violated. Some would take the Christian argument further. If the solution to the envisaged global food shortage implies widespread use of genetic engineering, it could imply that God had always positively intended human beings to make use of the genetic potential inherent in creation. On this basis, it might be argued that it is no more problematical than using the potential inherent in the earth's minerals to produce metals and alloys like iron, steel, bronze and aluminium. It could also be seen as being more 'natural' than the use of man-made toxic molecules to kill weeds, pests and diseases. On the other hand, a population explosion so great that conventional food supply methods are inadequate could suggest that something rather fundamental has gone wrong with regard to God's intentions. Transgenic food would then be regarded not as a *necessity* but a provision, in case of emergency, which God positively intended. Moreover, the unsustainable way in which minerals have been extracted, seen in retrospect, presents a warning not to exploit genes in a manner which makes similar mistakes. The commercial push for short-term returns on the genetic resource may not yield the optimum use of its benefits for all.

The Polkinghorne Committee discussed the general question of the unnaturalness of genetic modification involving food and the possible moral taint linked to it. They identified that those of the Islamic faith had inherent objections to genetic modifications on the grounds of the creation representing the best design through a Qur'anic basis of non-interference with the created world. The other groups they consulted did not appear to have inherent objections. The Polkinghorne Committee concluded 'that only a very small minority of the population would object in principle to all genetically modified foods on ethical grounds'. It must be said, however, that their sampling approach concentrated on groups representative of various communities and viewpoints of which the committee were aware, which is not the same as more general surveys of attitudes such as that carried out by Martin and Tait.[136]

Moreover, a lot of water has since flowed under this particular bridge, both in technical developments and in public mood. The concern expressed by such views as 'I'm worried because it's genetic' seems currently to be significant in the population at large. This may also have increased in the wake of the BSE crisis, *E. coli* outbreaks or sheep cloning. Even though none of these involved genetic engineering as such, the point often becomes confused and these all became lumped together into a general perception of technological intervention. Thus inherent objections to genetically modified food may now be a good deal more widespread than the Polkinghorne Committee concluded in 1993. As will be seen later, this leads to considerable concerns about how the issue is being currently being handled by society.

The Concept of 'Copy Genes'

It was observed above that transgenic food might present particular problems of principle for Jewish dietary regulations taken from the Old Testament, and the equivalent Qur'anic laws of Islam. This is particularly the case if an otherwise acceptable foodstuff now contains genetic material which originated in a ritually unclean animal, such as a pig. Will this mean that the food can not be eaten? Analogous considerations apply to Hindu, Sikh and Buddhist communities.[137]

In addition to these religious concerns, there are other groups and individuals in society with fundamental objections to eating food with foreign genes in general, or, in the case of vegetarians and vegans, food which might contain animal genes, such as the calf chymosin transposed into yeast. More generally in society, many people might baulk at the idea of foods which contained genes of human origin, seeing this as analogous to cannibalism.

This raises two important questions considered by the Polkinghorne Committee. Is it correct to say that the identity of the gene from the originating organism is transferred with it to another organism, or is a gene simply a complex chemical which is transferable across species without ethical reservations? For example, does recombinant DNA from a pig stay as pig DNA if put in a sheep, or is it just DNA? To what extent does the presence of genes foreign to the organism confer a moral taint on the food, such that it might indeed be cannibalism to eat food with a human gene in it?

In addressing these questions, the Polkinghorne Committee placed substantial intellectual emphasis upon what they termed the 'dilution effect', and introduced the concept of a 'copy gene'.

Consider an animal engineered to contain a gene of human origin. To obtain the human gene requires a process in which the human fragment of genetic material is copied millions of times in the laboratory, by inserting it into bacteria, which then rapidly multiply, or using other DNA copying techniques – such as polymerase chain reaction (PCR) (Figure 6.3, step 3). More detail of this process can be found in Chapter 1. It is these numerous *copies* of the original human gene that are then inserted into the animal, rather than genes that had actually been in the human body, hence the term copy genes. This process is thus said to have diluted the original human gene effectively to zero because one copy of the human DNA would have provided the large number of copies, one of which would have gone into one animal cell. There is also a disconnection due to the fact that most of the DNA inserted into the animal cells would have been produced by the bacterial intermediary or produced chemically through the PCR process.

The transgenic animal will thus contain almost no DNA molecules which had actually been inside the human body. Thus, it is argued, it does not carry the moral problem of containing human material. This distinction identified by the Polkinghorne Committee has been reiterated in many subsequent official publications.[138] This has proved to be an unfortunate precedent because the logic of this argument has by no means been universally accepted either amongst scientists or ethicists. It is also related to a point of deep controversy with regard to gene patenting (see Chapter 8).

To attempt to put the issue in perspective, the Polkinghorne Committee considered the ethical status of a hypothetical synthetic gene. Supposing a particular gene sequence was synthesised artificially from chemicals so that it was chemically identical to, say, a human gene, without ever having been in a human body. Surely, this would overcome the foreign gene objection completely, because the gene had literally never been part of a human? The Committee suggested that it would be unambiguously acceptable in food.

Despite this, the concept of dilution and the synthetic gene arguments seem to be of limited value ethically. Arguments based on the chemistry of transcription would not answer those whose inherent objections are based either on the fact that it was foreign to the receiving organism, or that the artificial transfer of inherent features from one organism to another amounts to tampering with nature or 'God's best design'.

A great deal depends on what is one's concept of identity with regard to a gene. For those for whom it is determined simply by where that particular gene has been synthesised, presumably the Polkinghorne argument would remove the ethical concern. A synthetic version of a gene of animal origin might satisfy some vegetarians, if the key thing was whether any molecule in the foodstuff had literally ever been in an animal, or had been obtained as a result of killing or obtaining blood or tissue from an animal. For those who take a less reductionist view, however, and consider that perceived identity is not just a matter of the history of one particular combination of carbon and other atoms, but something more associative, the argument is unconvincing. Many people perceive an association between a gene and its originating organism, no matter how many times and by what means it has been copied.

It is doubtful whether dilution answers the Committee's basic question. The DNA sequence within an organism represents the information required to produce a particular order of amino acids which form a protein. An objection could still be made to material made in a non-living system, or by multiple copies via a third party organism, on the basis of the distinctive nature of the genetic material itself and its initial place in creation, rather than to the way in which it was made. A gene is likely, at least to some extent, to be characteristic of its origin. Although most genetic material may be identical in several species, many genes will be an expression of a distinctive and perhaps even unique feature of the source organism, relating to its mutational or evolutionary history. In particular, the DNA conveys *information* which relates to the function of the gene in the source organism. Thus, in Case Study 8, sheep naturally produce a sheep version of the desired protein alpha–1–antitrypsin, but the human gene is introduced because it conveys the information to tell the sheep to produce the human version of the protein which is subtly, but vitally, different. It is the human-derived information that is the key to the whole application, not which species those particular atoms were last part of. To that extent, the concept of the dilution effect seems to miss the most important point.

The ultimate question in this area is whether this mixing conveys, in Polkinghorne's terminology, a moral taint. Clearly for some groups and individuals, whether by association, or by mixing genes at all, it matters enough to render some or all forms of genetically modified food unacceptable. For others, the primary identity of human, animal or plant is, as in a biblical view, that of the organism as a whole, and not merely the aggregate of its physical parts. A Jewish view given to the Polkinghorne Committee was to the effect that if it looks like a sheep, it's a sheep.[139] Thus the mere presence of one or two genes originating from a different species does not make a sheep less than a sheep, or irretrievably tainted with, say, pig or human, from a point of view of eating it.[140]

Ecological Objections

One of the loudest objections to genetic engineering in food crops has come from the ecological sphere, concerning its environmental and social consequences for world agriculture. To enhance crop yield or disease tolerance by the manipulation of a particular gene is the latest idea of a basically reductionist approach to science which ignores the interactive nature of ecosystems. Stress is put on the advantages in sustainability of agricultural methods which seek to exploit more fully the complex interactions and counterbalances of natural systems. It is argued that traditional mixed farming methods relied on these interactions, but that they have been disturbed or ignored in modern monocropping, even without genetically modified organisms

From the Old Testament's holistic view of life, agriculture has an ethical function which goes beyond human survival and the production of food. Through participating in its practices, people learn the value of a cooperative relationship with the land that goes beyond a mere utilitarian perspective. In this view, the purpose of agriculture is not just a matter of producing as much food per acre as is either physically possible or economically viable. It embraces other important goals, including the nurture of the landscape, the preservation of natural resources of soil, water and species for future generations, and the maintenance of healthy and viable rural communities. Whether this concept remains valid in a situation where only 2 to 3 per cent of the population are in contact with the land is a significant point for discussion. The question of transgenic food needs to be set in the context of this overall approach to agriculture.

It is significant that for all the benefits of industrialised societies, a lot of this perspective has been lost. Most people who live in cities or suburbs have little idea of how to raise crops or tend farm animals. The mass production metaphor, conceived in an urban factory environment making identical mechanical components, sits somewhat uncomfortably with a way of life dependent on cooperation with the natural world and our fellow creatures. There is a point beyond which the drive for efficiency begins to conflict with other values inherent in creation. While granting the importance of correcting past mistakes, some are sceptical of claims that biotechnology can solve problems which are deeper and more structural than can be addressed by technical fixes alone. Moreover, they suspect that remedial technology may create as many problems as it solves.

In reality, however, to construct a theme of conflict between these two approaches seems the wrong way to begin. Undoubtedly there are elements of ideological conflict, but to continue to stress this is not helpful if in reality both approaches may be needed, working in complementary ways. Some claims for ecological agriculture may be exaggerated, but at present the greater danger is probably that the push for more and more biotechnology is apt to ignore the less glamorous ecological aspects of agriculture, and the insights which more traditional forms can still offer.

CONSEQUENTIAL ISSUES

On the face of it, a range of considerable ethical benefits could be available from the developments in transgenic food listed earlier in this chapter. These need to be weighed up in the light of the sorts of underlying concerns described in the previous section, and also various disadvantages which can be identified. The environmental advantages of reducing chemical inputs to the land, the possibility of growing crops in hitherto marginal regions of the developing world, and the improvements in nutritional qualities of food are all fine goals as such, and would be welcomed by most people. If GMOs were going to make the difference between people going hungry or having enough to eat, then there would be a clear ethical case for the risks to ecological balance or human health to be worth taking. A number of questions need to be asked, however, before accepting the force of these arguments uncritically.

How Realistic are the Expectations?

Overselling an idea is a constant temptation for novel technology. Some grand claims are made of the potential, only some of which are likely to be fulfilled. The assertion that herbicide and pesticide inputs will be substantially reduced by transgenic crops remains to be widely demonstrated. There are some encouraging indications. For example, chemical inputs have been reduced through the use of genetically modified maize and cotton in US and Australia. Cotton engineered to express the natural *Bt* pesticide seems to require substantially less pesticide spraying,[141] but it has also been pointed out that fresh strategies will be needed if pests gradually become resistant.[142] It is too early to draw general conclusions about effectiveness and reduction of chemicals, especially as more and more applications of this type are being put forward for release. This will require rather careful and independent assessment, but there appear to be grounds for expecting some successes. Where genetically modified crops are tied to the use of particular chemicals already being marketed by the same firm, however, legitimate doubts have been voiced whether this is an optimum synergy for a long term trend to reduce chemical inputs to agriculture.

Not every application will prove feasible. In Chapter 1 reference was made to the idea of genetically engineering the nitrogen fixation abilities of soil bacteria into plants, to reduce or remove the need for nitrogenous fertilisers. Despite many years of work in both biochemistry and genetics, this has not proved possible. There are a number of factors causing this. One is that many different genes seem to be involved. A second factor may lie in the paradox that so few plants have evolved with a nitrogen fixing ability, when it ought at first sight to confer important advantages. A reason for this appears to be that to fix such an unreactive chemical as nitrogen directly from the air requires a considerable amount of energy, and requires a restructuring of the metabolism of the plant. It may have been more efficient to use the root system to draw ready-fixed nitrogen from the soil. Even if the genetics could be worked out, it is possible that the plants concerned would have metabolic disadvantages.

In general, genetic engineering has so far only worked well in applications where effects are controlled by only one or two genes. Much attention has been paid to herbicide, pest and pathogen resistance since these are often governed by single genes. To manipulate multiple gene effects is one of the future aims listed in Chapter 2. It is important to recognise that the direction of genetic engineering is not simply a question of choosing priorities from among endless possibilities. As a young technology it is inevitably shaped and limited by what is currently feasible.

It is thus important to have a realistic appreciation of the viability of what is claimed for plant applications of genetic engineering, and whether this will justify the work currently being put in. At present the technology is perhaps too young to see more than unbounded expectations. Once it moves out of this early phase, and a more sober estimate can be made, transgenesis may not be seen as a near universal panacea for food production. It will probably be one tool amongst many, powerful indeed in certain circumstances, but less useful or impractical in others. Those funding transgenic food research are also constantly assessing whether other aspects of agricultural research would yield a better return, not only for financial profit, but with regard to human need, animal welfare and environmental repair.

Health Risks from Genetically Modified Food

Despite its successes, the green revolution produced plenty of examples which were good neither for people nor nature. In applying novel technology, where reliable predictive data will by definition be scarce, both the environmental aspects and human risk need to be kept under review. Environmental risks are discussed more fully in Chapter 7, but the question of food safety is now considered briefly.

Human health and safety involves a number of aspects distinct from the issues discussed in relation to features of the production of transgenic organisms. One safety concern is the use of foreign protein coding genetic sequences which may cause allergic reactions in some people. A more prominent example which has raised public concern is the use of antibiotic resistance genes in transgenic food (see Figure 6.3, step 2a). Ciba-Geigy's genetically modified maize contains a gene conferring resistance to the antibiotic ampicillin. This is left over from the process used to modify the plant. In 1996, the UK Advisory Committee on Novel Foods and Processes did not recommend approval of this product because of a risk that the antibiotic resistance gene could be transferred into gut bacteria when the maize was used for animal feed in an unprocessed form. The probability was small and no precedent seems to exist, but the consequences could be large were it actually to occur. This was overruled by EU committees, because they apparently evaluated the risk to be lower. In December 1996, the EU duly allowed the maize to be introduced into Europe, but their decision was immediately challenged by some member states – Austria, Italy and Luxembourg defied the EU by banning the imports. At the time of writing this matter remains unresolved, but it is clear that the manner in which this and other decisions about genetically modified food have been made in Europe has provoked a public reaction.

It is beyond the scope of this work to assess these safety aspects, but the ethical issue here is that there should be a fair balance of risk between producers of the transgenic food, who stand to gain by it, and those who consume the food. It will be difficult to gain a consensus on what constitutes a fair balance, however. Industry, which will carry the costs of the evaluation of safety, would tend to wish to have a minimum system of controls. Emlay compared the relative merits of using a full toxicity test on transgenic food with an assessment based on a more limited number of attributes, aimed at demonstrating substantial equivalency.[143] His attributes suggested for assessment are:

- full characterisation of introduced DNA;

- assessing the stability of introduced genes;
- assessing the safety of the expressed products of the introduced genes, including toxicity and allergenicity;
- comparison of levels of significant nutritional components relative to traditional varietal bench-marks;
- comparison of levels of natural endogenous toxins with traditional varieties;
- confirmation of the absence of changes in agronomic characteristics, other than in relation to the desired effects.

Clearly this list would allow identification of most probable risks and it is based upon things which can be measured with confidence. It does not include the position of insertion of the introduced DNA, or the genes and sequences adjacent to the point of insertion. As Case Study 1 on *Pseudomonas* showed, these may be important in some circumstances. Whether more cursory risk assessment should now be allowed should clearly be a matter for wide and informed public discussion.

Who Needs Genetically Modified Food?

An impressive range of ethical advantages are claimed for genetically modified food. Not all these carry the same ethical weight, however. It is important to ask whether the claims are being borne out by the initial developments coming to market. It appears that these concentrate much less on areas which would confer substantial ethical benefits, and more on production characteristics or more cosmetic aspects, such as the taste, texture and appearance of food. This suggests it is necessary to look critically at what is actually being offered and why. At the beginning of this chapter, the question was posed whether genetically modified food is an indulgence for the rich or a necessity for the poor. A number of issues of justice underlie this, relating to who are the real beneficiaries and losers, whether the real human needs are being met or ignored, and whether the forces that drive the implementation of this technology are justly accountable to society.

Who are the Beneficiaries?

The first question to be asked is for whose benefit are genetically modified foods developed? Do they relate more to features of food production economics or to features of unique value to consumers? It is interesting to observe how the benefits are being presented to the public through the information leaflets of major UK supermarket chains. According to one, the benefits of genetically modified food are described as 'the potential it offers in terms of improved tastes and textures, reduction in waste and longer shelf life', but these benefits are not expressed in terms of 'the future of food production and farming'.[144] Another chain presented the benefits of genetically modified tomato puree primarily in terms of production convenience for Californian producers and reduced energy input.[145] The customer benefit is presented as reduced price. It implied this was a result of a lower unit cost, but it was not clear how much this was also to do with market promotion for a new product. The leaflet wisely did not make too strong a claim for something as subjective as improved flavour.

The driving forces would appear to be production characteristics, rather than human need as such or even customer preference. In general, improving production ought to be a good thing, but much depends on how its benefits are handed on and what wider consequences it may have. In the cases mentioned above, the balance of societal benefit seems to lie predominantly with large producers, and is not spread evenly over the whole population. In defence it is argued that improved production should lead to lower prices to the consumer, and that it would be sufficient justification in itself if genetically modified food is a means of increasing the overall economic prosperity of society. The problem with these broad economic justifications is that they tend simply to assume that there is a fair distribution of benefit, and that there are no social costs that would hit more vulnerable sections of society, whether at home or abroad. Neither of these is self-evidently the case. The linking of transgenic seeds to certain herbicides, for example, may make the small farmer too dependent on the seed company, so that technology becomes a burden rather than a liberation.

Production improvement also becomes more problematical as a justification where it involves an innovation which raises significant public misgivings. There is little justice in improving production characteristics if people are not convinced they want the product anyway. In this light a pertinent question is whether we really need non-squashy tomatoes (Case Study 6) or their paste equivalent. This does not appear to be an example of an important human need being met, but rather a reflection of the structure of the food supply system that has evolved in recent years. Reduced energy input was one of the main reasons given in favour of the modified puree. If energy saving was the driving concern, one might ask in the UK context how much could be saved by using more locally sourced tomatoes instead of shipping the product thousands of miles from California or even from Italy. The rationale assumes a transport intensive approach, made possible by low fuel costs, where food products can be sourced from far distant locations. There is increasing and widespread criticism of this on the grounds of the environmental and climatic damage involved.[146]

To the extent that environmental improvement is one of the key reasons for many transgenic developments, an alternative and logical approach would be a drive by the food production industry for a more regional system of supplying our supermarket shelves. The same argument could be put in criticism of a number of food quality and food production improvements. It is one thing where it is a question of supplementing the nutrition of a region in which a balanced and varied diet is very difficult to obtain in some seasons. It is quite another if we demand to override the seasons primarily for the sake of consumer convenience. Our present supermarket culture, driven by consumer wishes, needs now to weigh up whether this justifies the wider costs of which we are becoming aware.

Which Human Needs Should Genetic Engineering Focus On?

The next point concerns the opportunity cost of the types of genetically modified food currently being offered. To what extent could something better be done with the skills and resources devoted to such enterprises? If the claims made about the necessity of genetic engineering to feed the world are to be taken seriously, one may legitimately ask whether the effort and skill involved could have been better used addressing some pressing developing world food requirements. Given that currently there is not a problem with feeding the population of western Europe or the US, it is hard to see what justifies the priority for what are fairly cosmetic food developments for rich countries. Are we addressing the real problem or just finding increasingly sophisticated ways of feathering our nests while the rest of the world goes hungry?

There are two factors involved. One is the constraint that in a relatively new field, the technology is limited by what is available in the science. Of the various potential applications of genetics listed at the beginning of this chapter, not all are equally feasible or as far developed. It happened that tomatoes were relatively easy to study and that breakthroughs occurred in academic research for which applications were sought. This reflects something of the serendipitous nature of research that its successes cannot always be predicted. Areas which might be more useful may be more difficult technically. The nitrogen fixation example has already been given. Case Study 2 showed that the application to food crops modified to grow under conditions of abiotic stress – such as drought, cold and high mineral levels – also appears to be another situation where something which could give very large benefits may not be so easy. Important traits may be controlled by many genes and the species may be difficult to regenerate. In due course, solutions to some of these technical problems may well appear if genetically modified crops become more widely used, and experience grows.

The other factor is money. Pioneering genetic research is expensive and requires significant up-front investment. Only applications directed to the consumer markets of the world's rich nations are likely to offer the prospect of large paybacks. There are likely to be greater financial returns in pursuing western consumer preferences, than from meeting the needs of the poor. The former are likely to take precedence for the skills and investment in research.

Against these arguments, the claim for genetic engineering to feed the world continues to be made. If this claim is not to become meaningless, there will eventually need to be some rather concrete manifestations of how genetic engineering is going to meet the nutritional needs of two thirds of the world. Limited deployment of genetically modified crops is occurring in developing countries, but some very extensive releases have taken place in parts of China, holding a major proportion of the world population, with rather little regulatory control. Overall, however, present applications in genetically modified food do not encourage optimism in this respect. The emerging transgenic food industry thus has a question of morality and honesty to face. If it is to have credibility, it will need to show substantial investment in the less profitable business of feeding the poor. It has to show some major progress if the claim is to be believed.

Accountability in Decisions About Transgenic Food

The third question of justice concerns the extent to which the applications now being implemented have the support of the society concerned, and whether adequate means exist to determine and express public viewpoints. The situation in the UK in mid-1997 is not reassuring on either point.

The present transgenic cases of Monsanto's maize and Ciba-Geigy's soyabean both reveal disturbing trends. Given that something as basic as food and as contentious as genetic engineering is at issue, the amount of influence which private commercial concerns and international trade pressure can bring on public issues seems to be out of proportion. The fact that the companies concerned have not offered segregated supplies of modified and unmodified foodstuffs seems an example of aggressive attitude towards the public of another nation.[147] There are indications that supermarkets are seeking segregated supplies, in response to consumer pressure, but it should be asked why the situation arose in the first place.[148] The approval of genetically modified maize by the European

Commission in December 1996 without publishing their justification has led to much criticism and a refusal by some EU member states to comply, and a censure vote by the European Parliament.[149] The EU has since put pressure on Austria that its opposition was contrary to trade agreements but to date the matter has not been resolved. This has brought to light some disturbing questions. The European Commission maintains that it has no legislative competence to decide for member states on ethical matters on biotechnology, yet in this case it is apparently seeking to override the ethical stance of one of its member states. It has also been pointed out that once the US government authorised the maize and the producers decided not to segregate, the only way for Europe to have kept it out would have been to ban all US maize imports and thereby risk a major trade war.[150] In the face of such pressures the opinion of the people of Europe seems to be of only marginal importance. Monsanto has recently admitted that attitudes to biotechnology are very different between the EU and the US.[151] This raises serious questions of how far ethics are overridden by terms of trade, and citizens' just rights are made secondary to international legislation.

The driving forces behind the research into transgenic food, whether laudable or questionable, would appear to be far too remote from public involvement and democratic accountability. Since the end of the Second World War there has been a general trend away from public involvement with food production, with the result that few work in its production or have any say about what is offered. For most people in the UK, food is what is found on supermarket shelves, with little sense of connection with what it is or where it came from. Thus when any major change like genetic engineering is in prospect, it comes as an external factor to be explained to the public, who do not have a natural point of input, except at the very last stage. Public participation is apt to be reduced to the act of accepting or rejecting the packet on the shelf. In the view of some, that is the only ethical choice that counts, but this is an unnecessarily crude way for a society to make moral decisions. Where foodstuffs such as soyabean and maize are used in a such a huge range of different foodstuffs and not labelled as genetically engineered, it becomes very difficult, if not impossible, for the market to act as an ethical judge. It is also a very careless way of using its resources, if the public turn out not to want the product.

In conclusion, for an aspect of human life of such immediate concern to everyone, there is something wrong if the developments which the public are eventually offered are things over which they have had very little say in choosing. Greater public accountability is needed in this area, perhaps more than any other of the areas of genetic engineering considered in this study. At a point when the public is uncertain over whether it wants this new technology of not, it seems foolish to put on to the market genetically modified versions of the very types of foodstuffs which are so widely used in food processing that almost any food will contain them. The risk is that as with irradiated food, the public will vote with its feet about genetically modified food, and any benefits will be lost to everyone.

REGULATING TRANSGENIC FOODS

The Polkinghorne Committee

Concerns about the relationship between genetic engineering and food first arose over the non-transgenic sheep which had been part of a programme attempting to introduce a human gene for protein expression in milk (similar to those discussed in Case Study 8). By the nature of the techniques, most of the sheep used had not incorporated the gene, and were therefore not transgenic. Could these then be sold for food? Appreciating the seriousness of the issues and their impact on the population at large, in 1992 MAFF set up the Committee on the Ethics of Genetic Modification and Food Use, the Polkinghorne Committee referred to earlier. It reported its findings in 1993.[152]

Scope of the Investigation

The Polkinghorne Committee assessed the ethical implications of:

- transfer of human genes to food animals;
- transfer of genes from animal species which would be unacceptable as food for some religious groups, to 'acceptable' species;
- transfer of animal genes to food crops;
- use of organisms containing human genes as animal feeds.

The remit of the Committee is important to recognise. It did not consider the safety of foods produced by genetic modification, since this was the responsibility of the Advisory Committee on Novel Foods and Processes (ACNFP). More significantly it was not allowed to consider the ethics of plant or microbial genetic modification. Presumably this was because it was assumed that sensitivities would be less acute than those relating to animals. In retrospect this was probably a mistake. The Committee recognised but did not explore issues related to the nature of the technology, its consequences for the environment and issues related to the ownership or patenting of life forms, although they recognised that all of these raised ethical issues. The Polkinghorne Committee identified that there were issues which would be considered by some to induce a moral taint to the products of genetic modification and which, as such, would render such foods ethically unacceptable (that is, what gene combination is chosen in Figure 6.3, step 2). The principle of moral taint appeared to relate to the perception of the unnaturalness of the technology, and also to animal welfare.

Conclusions of the Polkinghorne Committee

The main conclusions of the Polkinghorne Committee were that:

* There was no overriding ethical objection to the use of organisms containing copy genes of human origin as food. They did, however, suggest that the use for food production of such genes, and other ethically sensitive copy genes such as those involving ritually unclean animals, should be discouraged where alternatives were available.
* Food products containing copy genes of human origin should be labelled to allow consumers to exercise choice, as should foods containing copy genes from animals which are the subject of religious dietary restrictions and plant products containing copy genes of animal origin which vegetarians might find unacceptable. These labelling recommendations did not, however, extend to food involving any other transgene, nor to a general requirement to label all genetically engineered foods.

Labelling: Meeting the Need of Those with Inherent Objections

From the ethical discussions earlier, it is clear that within our society there are individuals and groups who object inherently to any genetic modification involving food. This raises the question of what a democratic society should do about this situation, assuming it is a minority view. The extent to which this view should be accepted by others in UK society was excluded from the remit given to the Polkinghorne Committee, which thus did not make specific recommendations for this. Instead it only recommended labelling food involving copy genes of human or animal origin. The government has followed suit and did not demand the labelling of all transgenic foodstuffs. This indicates that they felt implicitly that inherent objections should be addressed only if they were based upon certain criteria, which they defined as movement of human genes to other species, or of animal genes to plants, but not if they involved other modifications such as the transfer of bacterial genes into plants.

In 1995, the ACNFP agreed that three processed foods derived from genetically modified plants could be cleared for sale in Britain. These were a tomato paste from genetically modified tomatoes, oil from genetically modified oilseed rape and processed products from soyabean. The Committee argued that because the new products were almost identical to existing products, save for the specific agreed modifications, the foods required no specific labelling. It did, however, encourage firms to develop 'informative labelling on a voluntary basis'. It is significant that supermarket chains have considered it prudent to label the genetically modified tomato paste and are taking steps over the case of soyabean and maize to demand segregated sources where universal labelling of the products would be very difficult. In response to the controversy following its decision in December to allow geneticaly modifed maize into Europe, the European Commission has been trying to find a system for mandatory labelling of certain types of modified food. It proposed in December 1997 that foods containing genetically modifed soya and maize should be labelled if protein tests could detect the presence of the new genes or protein, but agreement has not so far been possible over the terms of the extent of the coverage or the wording.[153]

These developments were the first in the UK involving transgenic foods, and so may be seen as setting a precedent. They reveal a number of disturbing features. Inherent objections to genetic modifications appear not to have been considered, because they could not be fitted into an evaluation framework and thus have become somewhat marginalised. Since the criterion for labelling was not based on general intrinsic objections, but only on whether the products to be released contained genes of human or animal origin, no labels were required for any of the above products. This meant that those with inherent objections were unable to avoid these products, and those providing basic foods for processing or for consumption outwith the home would not be able to provide information on source. Indeed, no recommendations at all were made in relation to labelling of products sold on into restaurants and other public places. Finally, as discussed above, the rationale for the modification was related to improved production features rather than to features of unique value to consumers. The balance of benefit from the modification was not spread evenly over the whole population, but lay predominantly with large producers.

Inadequacies in the Present System for Regulating Transgenic Foods

The manner in which the Polkinghorne recommendations have become interpreted in subsequent official and industry documents is a cause of some concern. In 1995, MAFF issued guidelines for the public on genetic modification of food.[154] This document followed the line of reasoning of the Polkinghorne Committee and used its report as the answer to any and all ethical questions, regardless of the restricted remit of the Polkinghorne Committee. The information in the MAFF document is further diluted in the food industry document *Food for our Future*[155] and in information released from supermarket chains.

The MAFF document *Genetic Modification of Food* justifies the acceptability of moving genes between plants and animals which do not normally inter-breed, by reference to the establishment of the Polkinghorne Committee. The Polkinghorne Committee, however, explicitly indicated that '… neither were we asked to consider the wider question of the ethics of genetic modification in itself'. Similarly, *Food for our Future* asks the question 'Is it ethical?' and states 'The UK government set up a Committee to consider these points, chaired by Professor Polkinghorne.' It goes on to summarise the conclusions, such as 'Most Christian and Jewish groups in general find genetic modification acceptable'. This is stated despite the fact that the Committee did not in fact draw this particular conclusion, but indicated that significant reservations were expressed over some kinds of modification. The report of the UK National Consensus Conference on Plant Biotechnology (1994)[156] included among its conclusions 'Finally, since the technology is in place and scientific knowledge of biotechnology is increasing daily, the ultimate choice lies with the consumer who should be adequately informed of what he or she may choose'. This does not seem to sit easily with current discussion on the labelling of only some transgenic foods.

Society has always had difficulty in meeting the needs of minorities with inherent objections on major issues in public life. Traditionally, for substantial minorities, this type of problem has been covered by allowing the minority to avoid it using the labelling approach. Very small minorities and their needs and views tend to be ignored, it being simply impractical to do otherwise. This would appear to be the stance taken in respect of transgenic foods. Given the central nature of food in society, it is manifestly unethical for society not to consider the views of those who have inherent objections to genetic engineering. There is a clear moral obligation for society to provide them with the means of exercising choice, and without imposing additional cost on them of a change from their normal eating patterns. The critical problem, however, is not foods which are easily identified and labelled such as a tomato, but the foodstuffs like soyabean and maize that are used so widely in food processing that almost any food may contain them. In these cases, it would seem that steps must be taken to ensure that public concerns are not overridden, and that sources can be readily identified by the consumer.

The relevant UK supermarkets selling genetically modified produce have put considerable energy into label design and into providing backup information, although it is not clear how much of this is reaching the average customer. It is recognised that labelling is not as straightforward as putting on a label saying 'Produced Using Genetic Engineering'. It has often been pointed out that the majority of people are fairly ignorant about what is meant by what is on the label. For example, most people do not seem to know that 'Farm Fresh Eggs' can come from battery cages or what an E number is. It is also reasonable to ask what other production information should be included, and where one draws the line. It is thus likely to be difficult for labelling to incorporate adequate and balanced information. While this may be the case, it is not acceptable as an excuse for not trying.

There is also a problem of connotations. Some proponents of genetic modification argue that to label a food as genetically modified has a pejorative effect on the product, in that to be labelling it implies there is something wrong with it. They suggest that many people would avoid such products if it was drawn to their attention that they were genetically modified. This is by no means proven. The connotation of a label depends on how the society at large views the product, and this can go either way. Thus when Flavr Savr™ tomatoes were labelled 'Product of Genetic Engineering' in the US, there is some evidence that sales increased. On the other hand, if for most people genetic engineering remains something mysterious, there would probably be some truth in the pejorative notion, but even this is better than leaving everyone ignorant over the issue. It matters to people that they are given information they would wish for, and that they can trust the source of the information. If public discussion of the issue becomes wider, one may hope people will be able to make up their own minds. It would seem that the biotechnology industry has every-thing to gain from genetically modified food labelling, quite apart from the clear moral imperative to do so. To gain acceptance for such products by deception or failing to disclose such information would be a hollow victory, which might well backfire in the long run.

Conclusions

Transgenic food has emerged as a major issue of public concern. In so rapidly moving an area, however, it is not possible to answer every aspect of the many ethical questions involved. This chapter has reflected on overall principles such as whether it is acceptable to use genetic modification for foodstuffs. The case for an inherent objection was not ultimately convincing for most of our group, but it is most important to recognise and provide for the many who will reject such foods on ethical, environmental or religious grounds. The argument that copy genes lose any sense of identification with the organism from which they were derived was not found persuasive to most of the group. Legitimate questions appear to exist over what effect transgenic foods will have on existing concerns over intensive agriculture, which is possibly too dominated by the production motive for its own or society's good. If, however, the claims for substantially reducing chemical inputs can be converted into practice consistently, there would be important environmental advantages for some applications of the technology.

Attention was given to some specific cases which are important at the present time, because they seem to raise more general questions. These relate to the investment priorities in this area, the influence of private interests on what is offered for us to eat, the adequacy of public accountability for decisions in this area, and the need for labelling of transgenic foods. The central position of food in society suggests that the regulatory system in relation to transgenic foods should, as a minimum, allow those with inherent objections to avoid such products. There is a question of justice which seems to be compromised by the present situation.

Existing government committees, such as the ACNFP, have been constituted with the aim of monitoring supply issues relating to transgenic foods in a similar way to their mechanisms for assessing other safety issues. Developments in relation to issues such as BSE indicate that expert committees of this type do not always get it right. The consequences of error in respect of food are probably greater than for many other areas of genetic engineering. Ethical issues, especially those relating to inherent objections, do not appear to be regularly monitored, although the Banner Committee[157] called for the establishment of a group with this function. Although the Polkinghorne Committee reported as recently as 1993, the limitations of its specific terms of reference and the above points suggest that there are clear reasons to revisit the issue of transgenic foods.

The presence of transgenic foods on the market suggests that public debate on the issues raised here is needed now before the political agenda is carried too far either by opponents or proponents. There are clearly many potential ethical benefits in this area, not least to developing countries. It remains open to question whether these benefits will be realised, or whether the continuing direction of this field will be to follow the more trivial examples with which transgenic food seems to have begun.

7 LETTING OUT THE GENIE: ENVIRONMENTAL RISK AND REGULATION

INTRODUCTION

The element of risk is undoubtedly one of the most important themes in the public debate over biotechnology and genetic engineering. The power inherent in the tools of genetic manipulation calls for especial levels of skill, care and foresight, so that the planned applications do not also result in serious unintended consequences. A number of the case studies have identified potential risks associated with the use of GMOs in agriculture and food production. It would be unfair to claim that these risks have been ignored. Ever since scientists were first able to alter the genetic composition of organisms in ways which could not have arisen naturally, there have been attempts to ensure that these new technological skills would not give rise to human or environmental hazards. This has continued to be an area of widespread concern, however. The point in debate is whether the response to the potential risks has been sufficient in this relatively new area of technology, where there are inevitably many unknowns, and especially where perceptions of risk differ greatly within society. There is a dichotomy between the opinions of laboratory scientists and ecologists as to the likelihood of these possible problems occurring in practice, but at this stage in the development of the technology there is little concrete evidence on which to base statistical predictions.

This chapter first summarises the nature of the potential hazards from releasing GMOs into the environment, and then considers the different ways the risk is perceived. It examines the official response in the UK to these hazards, which was to adopt a precautionary approach to regulation. This marked a major departure from the previous reactive approach to risk regulation, which is based on predicting possible risks on the basis of known effects. It has led to subsequent debate over how far the precautionary approach can or should be relaxed. This raises a number of ethical issues concerning how a society should treat risks from incompletely understood phenomena and novel technological proposals. The primary focus is on environmental risk. In general, direct human health and safety risks are beyond the scope of this study, although the special case of disease transmission in xenotransplantation is discussed briefly in Chapter 5.

NATURE OF THE HAZARDS ARISING FROM THE RELEASE OF GMOS

Containment, Release and Monitoring

Most basic research into genetic engineering in plants and micro-organisms is done under what are termed 'containment conditions'. This means a combination of engineered systems and administrative procedures which are designed to contain the modified organism within strict physical bounds, in a laboratory setting. Eventually a promising area of research would need to be moved out of the laboratory, in order to be tested in a more normal environmental setting, in the form of a field trial. Typically, a new strain of genetically modified plants is sown in a field and monitored over a period of time against a range of risks. As well as observing the desired traits of the crop, its relationship with the wider environment is also examined.

With chemical pesticides, it is possible to use the trial stage to ensure that new products do not have any dangerous attributes. Any product which demonstrates potential to cause harm is not cleared for use, thus avoiding any problems. However, when a living organism is being tested, there is always a chance that even a trial release could lead to an uncontrollable situation should the organism prove capable of reproducing and spreading in the environment. An example of such a situation has arisen in Australia where a non-genetically manipulated calicivirus was being tested as a potential killing agent for rabbits in the wild, under strict quarantine on an offshore island. After a short period of time, however, the disease was found on the mainland, possibly transferred by an insect vector or by birds, where it rapidly spread out of control among the wild rabbit population with unpredictable consequences.[158]

Potential Hazards Which Might Arise From GMO Release

Among the factors to be considered are the spread of material beyond the physical limits of the field, and its dispersion beyond the genetic bounds of the species to neighbouring plants, both related and unrelated. In some cases effects on local ecosystems and biodiversity may be important, for example, where an unusual species is being introduced, or where an existing species like a weed is being challenged. Where the crop is not a food product, potential pathways into the food chain of both humans and animals have to be assessed, for example honey produced from the pollen of transgenic plants like oilseed rape (Case Study 3). If the genetic modification is directed towards pest or disease control, it is necessary to examine the impact of the control mechanism on other species – whether creating a toxic effect or conferring resistance. The effect of the scorpion toxin produced by the modified NP virus described in Case Study 5 is one example. There may also be unintended side effects due to effects at the site of insertion of the new gene, or from the use of an antibiotic resistance gene as a marker, as in the production of transgenic tomatoes (Case Study 6) or BST (Case Study 7).

There are particular concerns about the unintended spread of the introduced gene – either into the wild population of the plant, or into weed species, which might then produce more insect, disease or herbicide resistant forms of the weeds. There are various mechanisms by which this could occur. One is via 'volunteer' crops. For all that one year's genetically modified crop may have remained well contained, it may be impossible to avoid some plants reappearing in that field in the following year. This in effect constitutes a form of escape, if the field is now sown with a different crop.

Genetic manipulation of farm animals is not usually regarded as being potentially hazardous to the environment because in general they are more controllable than plants or micro-organisms. Genetic manipulation of farmed fish is, however, a significant exception, because escapes from fish cages occur frequently, and it would be impossible to prevent contamination of the wild gene pool with the genetically engineered traits. Two types of genetic manipulation of fish are giving rise to concern. One is the insertion of a gene leading to increased levels of production of growth hormone, and hence more rapid growth to a larger size. The other is the insertion of a trout gene into salmon smolts to improve their survival rate on transfer from fresh to salt water. It is reasoned that the ability to make this transfer more readily could lead to major changes in the behaviour of any such salmon which escaped from cages and that such behavioural traits may give these salmon a selective advantage which could then be passed on to other wild salmon.

Risks Identified by the Royal Commission on Environmental Pollution

One of the most important UK studies on the subject was the report of the Royal Commission on Environmental Pollution (RCEP), *The Release of Genetically Engineered Organisms into the Environment.*[159] It was published in 1989, during the period when the Department of the Environment was drafting the legislation to control the release of GMOs.[160] The Commission advocated a precautionary approach to regulation, taking an ecological perspective. They concluded that:

> *Organisms derived by genetic engineering can contain genetic information and exhibit properties that have evolved in the context of an unrelated species. These organisms may be produced in days or weeks, rather than the years required for traditional breeding techniques or the millennia for evolution. They are products of the laboratory and may well contain genes that are extremely unlikely to have occurred in nature in situations where the organism in question could multiply.*

The RCEP illustrated a general difficulty of predicting which organisms are likely to acquire pest status by the example of an obscure and rare British grass, *Vulpia vasciculata*, which has become a major introduced weed in Australia. Such a change of status could not have been predicted by a detailed study of its ecology in Britain. There is a critical threshold density below which the population of any species will probably fail to survive, no matter what is done to try to save them. Above this level, long term survival is much more likely. This difference is important when information from a trial release is being extrapolated to predict the outcome of repeated, large scale applications of a commercial product.[161]

The RCEP examined the possibility of transfer of genetic material from a GMO to another organism. This might occur for example by plasmid transfer between micro-organisms or via pollen dispersal. This is particularly relevant to the development of herbicide resistant plants where the trait of herbicide resistance may be transferred to neighbouring weeds. Some plants such as oilseed rape and oats already contain genes necessary for viability as weeds, and so relatively small changes in their genetic makeup could lead to their becoming weeds in practice. The same may be true of many other wild plants whose ecology is less well understood. Problems could also arise with the insertion of insecticidal genes into plants or micro-organisms. If these are spread to other non-commercial species, they could affect insects which are not pests, and which may even be protected species.

It is sometimes claimed that plant modifications made by deleting genes should be considered safe as such GMOs will be less fitted for survival than naturally occurring counterparts. The RCEP report was sceptical of such claims, since a deletion can profoundly alter the behaviour of an organism and, after release, the GMOs themselves may adapt in response to selection pressure.

Based on such considerations, the RCEP suggested that the following scientific information should be supplied to any panel set up to grant permission for the release of GMOs:

- a description of the parent organism, any vector and the resultant GMO, including relevant biological and ecological information;
- a description of the manipulation to produce the GMO, including its possible unwanted effects;
- the potential environmental effects, including information on any previous related releases;
- the objectives of the release; and
- the location of the proposed release, including relevant geographic and environmental information.

The above are examples of direct risks, that is those considered to be intrinsic to the product itself. In so far as these risks have been identified and analysed, and provided an appropriate regulatory system exists and functions as intended, the degree of risk should be minimised.

The RCEP also considered the indirect risks which could occur from the release of GMOs. Such risks can arise as a result of the way people treat or make use of the product, through carelessness, failure to observe regulations or misusing the products. For example, the staff in charge of a trial release may fail to observe recommended precautions. Once a product is available commercially, a farmer may mix two or more GMOs, increasing the chance of transfer of genetic material with unpredictable results.

To minimise the likelihood of such risks, the RCEP suggested that any regulatory system should include a requirement for information about the fitness of anyone applying to make a GMO release:

- the identity of personnel involved, including their qualifications and training;
- the arrangements for the release, including the preparation of the site, the timing of the release, and any subsequent dismantling or decontamination of the site;
- the monitoring arrangements;
- the contingency plans in case of unexpected events;
- the results of a prior local assessment and consultation.

Varying Perceptions of Novelty and Risk

It is one thing to identify a risk and to assess it in scientific terms. It is quite another how that risk is perceived outside the scientific community. This is a generic phenomenon, which first became prominent over nuclear power, but applies to many areas of technology. In the case of genetically modified organisms, the risks can be perceived in different ways depending on what they are likened to. As discussed above, an environmentalist might point to instances of plants introduced from other countries which subsequently became pests in the new habitat. If scientists cannot anticipate the consequences of conventional movements of genetic material around the planet, it is argued, how can they hope to predict with accuracy the possible deleterious consequences of the new genetically modified strains? Indeed the very use of the word release already raises concerns that we are letting something out which should not be let out. Alternatively, an advocate for the technology might stress how little genetic material is usually found to spread, that tight regulations exist, and of course the advantages that would result from the enterprise. This encapsulates two different perspectives of risk, a situation sometimes likened to the description of a glass of water as half full or half empty. The significance lies in the different stress on what is important.

Such different stresses are not only between scientists and public. Entrepreneurs working in the biotechnology industry will, when they are attempting to raise funding for some new venture or especially in citing grounds for novelty when applying for a patent, stress how different it is from what has gone before. The talk is in terms of revolution, of how biotechnology will enable us to gain unprecedented levels of control over living matter, to reprogram the computer software of the cell and to create new forms of organism that could not arise by natural means, and so on.[162] When the same entrepreneurs are discussing the need for regulation of this emerging industry, they are more likely to claim that biotechnology is really not a new science at all, and how little is really being changed. Here they talk of the parallels with the baking of bread, the brewing of beer and the selective breeding of farm animals and crops, and how these have progressed for centuries without causing direct environmental damage. Biotechnology is viewed as merely improving the precision and the speed with which we can continue to make progress.[163]

A dichotomy also exists in the scientific community. Laboratory-based scientists generally claim that the risks are severely overstated. Any genetically engineered organism is bound to be less fit than its wild counterpart and will not be able to survive and establish itself permanently in the environment, and therefore the risks of environmental damage arising from a release of a GMO are insignificant.[164] A precautionary approach is seen by such scientists as an unnecessary and wasteful restraint on technological innovation.

From the perspective of the ecological scientist, the picture is less reassuring. Ecologists tend to regard the introduction of non-native species into an ecosystem as the most appropriate analogy for the release of a GMO. They point to the difficulty of predicting which species will fail to establish themselves and which will become pests. Where species do become pests, whether they had been introduced deliberately or accidentally, ecologists point to the difficulty of regaining control, citing recent UK examples like Dutch elm disease and the spread of rhododendron.

A comprehensive review by a group of US ecologists[165] suggests that a precautionary scientific scrutiny of any proposed environmental release should be carried out, covering the following points:

- the nature of the genetic alteration, eg low genetic stability;
- the nature of the parent organism, eg whether it has pest status;
- the attributes of the engineered organism, eg whether it has enhanced fitness with respect to survival in the environment; and
- aspects of the environment itself, eg the presence of selection pressure for the engineered trait.

Where an organism or the relevant environment exhibits any problematic attributes, these ecologists feel that a high level of scientific scrutiny should be required, triggering a need for regulation during the trial release stages of product development.

The differences of perspective among laboratory scientists and ecologists are also reflected in the views of individuals involved in the regulatory system, some of whom see themselves as helping to prevent the occurrence of real risks to the environment, and some of whom see the primary purpose of the regulatory system as being to provide public reassurance about the safety of GMOs.[166] For example, in 1997 English

Nature and similar UK government agencies called for a three-year moratorium on herbicide resistant GMO crop releases particularly while further work is done on the spread of genes to wild species and the diversity implications for local ecosystems.

To summarise the views of the two sides of the risk debate presented in this section, industry and laboratory scientists tend to see GMOs as analogous to other field crops or farm animals, or to the micro-organisms used to bake bread or brew yeast, familiar, predictable and very unlikely to get out of control. Ecologists, on the other hand, see a closer analogy with GMOs in the various introduced species that have often created uncontrollable havoc in ecosystems different from those where they evolved naturally. This, however, represents the debate in its most specialised, scientific form, amongst those well acquainted with the issues. In the wider public arena, where the degree of knowledge is inevitably much less and where perceptions are subject to many different influences, things are more complex still, as discussed in Chapter 10.

DIFFERENT APPROACHES TO RISK REGULATION

So far there have been few instances where a GMO has shown unexpected, damaging side effects or has got out of control. This could be ascribed to the inherent safety of genetic technology, or alternatively it could be attributed to the success of the precautionary systems of risk regulation set up in the UK and the EU. Before considering the wider ethical implications, it is necessary to examine the two different concepts underlying the regulatory system, and the historical context in which these have arisen.

The Reactive/Preventive Approach

The 1950s and 1960s saw sweeping changes in the systems used to produce our food, sometimes referred to as the green revolution. Urged on by government concerns to avoid food shortages and backed up by a wide range of subsidies, the agrichemical industry produced the pesticides and fertilisers that enabled farmers to achieve increasing levels of productivity from their land.

One unfortunate side effect of this revolution was the damage inflicted upon wildlife, particularly by the organochlorine insecticides such as dichloro-diphenyl-trichloro-ethane (DDT). The environmental persistence of these chemicals was initially seen as a positive advantage since it prolonged the period over which they remained active on the crop. After a period of years, however, and as the use of these chemicals increased, it became apparent that their persistence had some disadvantages. Their levels in animal tissues began to build up as they were passed along various food webs until they reached lethal quantities in predators at the top of the food chain. In many parts of the world, peregrine falcons, golden eagles and other such species began to disappear rapidly.

The agrichemical companies involved in the production of these chemicals were initially reluctant to accept the evidence of the damage being caused. They demanded very convincing statistical standards of proof of damage before they would accept action being taken to limit the use of organochlorine insecticides. A public distrust of agrichemicals was stimulated by these events, which still affects the industry's image today.

Eventually it was accepted that undue persistence of a chemical in the environment could have serious consequences. Regulatory systems were developed to ensure that new pesticides did not give rise to risks of a similar nature. From that point on, any chemical that was not broken down by natural processes, such as the action of sunlight or microbial action, would be eliminated by the agrichemical industry at an early stage in the research and development process.

Unexpectedly, subsequent generations of pesticides showed different types of negative environmental impact. For example, some insecticidal seed dressings showed a selective toxicity towards geese.[167] As a result, the regulatory system gradually evolved to deal with each new risk, once its existence had been established by standard statistical procedures.

This is known as the reactive or preventive approach to risk regulation.[168] It is typical of the systems in place in the UK for dealing with most of the risks that arise from modern technology. The industry concerned and its products are controlled by a system set up in response to statistically proven adverse impacts that have arisen in earlier generations of products. Often these are brought to light only after one or more serious incidents or fatal accidents. Thereafter, new products and processes are screened to ensure that they do not give rise to any hazards similar to those which have happened before. By its nature, this type of regulatory system is built up slowly, in a piecemeal fashion, as new generations of product or process exhibit different hazards.

Another feature is that decisions about the need for regulation and the level of regulation required are taken in relation to the relevant benefits to society, compared to the costs to industry. For example, when the fungicides benomyl and thiophanate methyl were found to be highly toxic to earthworms,[169] it was decided that their potential benefits to agriculture outweighed the likely costs arising from this side effect and no restriction was placed on their use.

Reactive regulatory systems of this nature are the foundation of the control of the development of drugs, pesticides and other chemicals. They are also the basis for most systems of pollution control and health and safety legislation. The fact that these systems incorporate predictive tests for toxicity, for any tendency to cause genetic mutation or birth defects, and for environmental impacts does not disguise their essentially reactive nature in that they do not consider novel hazards before there is any direct evidence for their existence.

Precautionary Regulation of Genetically Modified Organisms

As a result of some of the negative experiences arising from earlier techno-logical developments, a new, proactive approach, known as the precautionary principle, began to emerge in Germany.[170] Although features of this approach can be seen, for example, in some elements of nuclear risk assessment, it was biotechnology, and particularly genetic manipulation, which was the first area where a systematic attempt was made to apply the precautionary principle to an emerging industry. To quote from the RCEP report, 'the opportunity exists to learn from the experiences and the predictions of the past in order to build environ-mental foresight into any necessary regulation of these new products'.[171]

A precautionary approach was applied to the biotechnology industry at the very earliest stages of its development. As soon as it became clear that the creation of recombinant DNA was technically possible, a world-wide moratorium on this area of research was agreed, at a conference held in Asilomar in California in 1975. This was a voluntary agreement among scientists, who were mainly working in universities and govern-ment research laboratories, to allow time to assess the potential hazards arising from the new techniques.[172] At that time these were mainly consid-ered to be hazards to human health. Following these debates, methods were devised for assessing the likely hazards from the release of different types of organism. Regulatory procedures were set up, to keep those which were deemed hazardous under contained conditions throughout all stages of an experiment or trial, both at the laboratory and industrial scales. In the light of these developments, the moratorium was relaxed over several years,

In the agricultural context, the main focus of attention in precaution-ary biotechnology regulation has moved to the control of products which are designed to be released live into the environment, rather than those which may accidentally escape from contained conditions.

In a precautionary approach, the industry and its products are controlled by a system set up to avoid potential hazards, predicted *in advance* of the development or the marketing of products. It is done before there is any empirical evidence of the existence of such hazards. In such a situation, safety issues and the formulation of the regulatory system become an integral part of the development of the industry itself from its outset.

Implementing a precautionary approach to regulation in this way, ahead of scientific proof of actual damage, presents major challenges to the scientific-rationalist worldview of the risk regulator. There are difficulties in attempting to predict the nature and extent of future hazards, or the likelihood of their occurrence. The human imagination can be boundlessly inventive, but we can also fail to foresee outcomes that are, with hindsight, blatantly obvious. An imagination constrained either by self interest or by prejudice could lead to serious environmental damage and a resultant loss of public trust in both industry and regulators. Equally, an overactive imagination could lead to needless restrictions on industrial innovation, which could deprive us of potentially useful products.

When Should a Precautionary Approach be Adopted?

From the point of view of the wider society, an important responsibility must fall on the commercial producer to ensure that due care has been taken to assess risks that might arise from a new product. The question is what entails a sufficient discharge of this responsibility. From the producer's perspective, a precautionary approach may impose extra development costs and may mean delays in bringing the products to the market place. For these reasons there are strong commercial and political pressures for the traditional reactive or preventive method of risk regulation to be the norm. It is argued that the more demanding precautionary approach can only be justified if there is a sufficient level of either complexity or uncertainty, or both.

In the case of GMOs, complexity is an important factor. The reactive regulatory approach relies on being able to analyse the effects of the different phenomena involved. By their nature, the complexity or scale of a set of environmental interactions may make it impossible to identify the cause and effect relationships on which the reactive approach depends. For example, in a case where large numbers of pollutants or organisms interact with one another, it may be impossible to identify the actual causes of environmental harm. Soil processes represent another inherently complex system.

With very new technologies, uncertainty plays a large role, but it may only be temporary. Where there is no previous experience on which to predict their impact on ecosystems, a precautionary approach may be justified until such time as a large and reassuring body of data has been accumulated. Once established, typically the technology begins to be treated as more familiar, and arguments are then advanced for taking less precaution. The moot point is when, if ever, it is responsible to declare that such a stage has been reached. Opinions on this point will depend, among other things, on the attitudes and prejudices of the decision makers.

Where it is agreed that GMOs require a precautionary approach, this is usually justified on grounds of uncertainty. It is assumed that, as we gain experience with the new technology and provided this experience is not negative, we will be able to relax our precautionary vigilance. However, as more products become available on the market and as they are increasingly widely used, the range and complexity of potential interactions between organisms will increase rapidly and this may justify maintaining a precautionary approach.

An example comes from a report prepared for the Council of Europe[173] which looked at the long term impact of GMOs. It emphasised the need for a precautionary approach, and included a comprehensive literature review. A survey of existing risk assessment procedures showed that emphasis is placed on the early stages immediately following a release, but that there was insufficient focus on long term ecological effects. The report suggested an approach to these long term effects which could be used to augment the current procedures. The approach is based on a systematic analysis of potential ecological interaction, using case studies of known examples of introductions of non-modified organisms.

ETHICAL ISSUES RELATED TO RISK FROM GENETIC ENGINEERING

Two Ethical Problems

This chapter has examined the nature of the risks from genetically modified organisms, the different ways these are perceived, and the way the regulatory system expresses an institutional response. The conflicts which have arisen are a reflection of an underlying ethical dilemma. Two ethical issues and their corollaries seem to stand out.

The first concerns the nature of our knowledge, with regard to our responsibilities. It asks whether we are actually in a position to determine adequately how far the developments are affecting the environment or human health, to know whether or not our intervention will turn out to be harmful. The corollary to this is the question of what represents adequacy of information and prediction, and who decides it.

The second concerns the social application of risk and its distribution. In pursuing a particular programme of genetically modified plants, are particular groups in society subjected to an unjust share of the burden of any risk? This includes concern for future generations and other countries. By the same token the distribution of benefits may pose an ethical problem, if they are primarily, or even exclusively, for certain groups in society, rather than shared out generally to most people.

Reactive and Precautionary Ethics

A conventional assessment of the ethical dimension in risk regulation is usually confined to the relative distribution of risks and benefits. A major difficulty with this lies in the highly variable thresholds of acceptability of risks, in relation to benefits, as perceived by different groups or individuals. Instead of this, a more instructive method is to examine how the two approaches to risk regulation present different analyses to these above questions, each expressing the underlying ethical assumptions which that approach embodies.

The reactive approach is based on the presupposition that the only risks a society should have to regulate are those it can reasonably foresee as a result of past evidence. Technology is seen as a fundamentally desirable thing for society and should be given its head. Its expected benefits to society as a whole – in economics, employment, quality of life – are deemed greater than the risks incurred overall, and greater than the higher risks certain individuals and groups are likely to suffer. It is thus a largely utilitarian view. The greater good of the many justifies the risks that the few may bear. We should only respond to what it is rational to respond to. Thus we must make the best estimate we can in the light of the knowledge we have, and proceed accordingly. It is ethically wrong to hold back from bringing in those benefits through unnecessary fears, provided society has acted on the best information it has. The driving force of this ethical view borrows much from the Enlightenment project of human mastery over nature through our ability to reason, and the optimistic notion of progress through scientific and technological advancement. The importance of this as an ethic in itself is missed by many people who subscribe to it.

The precautionary approach is also derived from experience, but it interprets it in a different way. It focuses on the fact that past events teach us that we can never be certain of the outcome of technology. Things nearly always go wrong that no one anticipated at the start. It reasons that human beings have only a limited right to subject society to these unknown, or at least ill-defined, risks and dangers. If these are big enough, we should try to anticipate them in legislation. In the limit, we should not take those risks at all unless and until we judge that we are justified in doing so. Above all, we should not wait until something has gone wrong, for which there are then real scientific data to assess. By that time, there will already have been unnecessary problems for people or the environment which could have been avoided if the right precautions had been taken first. The approach is capable of relaxing its precaution where experience dictates, but always recognises the element of unknowing which, more than our limited ability to quantify, characterises human existence.

The precautionary approach draws on Christian ethical insights, as well as on feminist and ecological perspectives, with a focus on the connectedness of the way things are. It puts a greater emphasis on goals and imperatives other than the economic or scientific. It espouses an ethic of cooperation towards the created order, rather than of intervening in order to dominate. It also incorporates a sense of humility which recognises both the finiteness and fallibility of humanity, in contrast to the hubris that has sometimes disfigured the progress motif. It also highlights the plight of the groups most vulnerable to the particular risks, setting greater store by the need to protect them than by the broader benefits to society, and setting a higher value on the longer term future than on the present or immediate future.

Either approach can have its problems. If the precautionary approach is taken too far, it loses something which many Christians would regard as God-given in an element of risk taking about life. Entering into a human relationship entails risk, as does a young person leaving home and making a life for themselves. It also encourages a romantic attitude to the here and now. This can play down present troubles like disease and suffering, which human endeavour might be able to help, but never without an element of risk.

The reactive approach relies too much on what it thinks it knows, and especially what it thinks it can measure. The limitations of knowledge lay it wide open to failure if it allows no room for an element of scepticism about its judgements. In common with most utilitarian thinking, it also has an inadequate means for providing justice for the few. Also, most systems which rely on specialist knowledge run the risk that the broader public, whom they purport to serve, does not identify with the 'expert' decisions being made on its behalf, and instead prefers less scientific perceptions.

Being Cautious in How Far to Apply Precaution?

In the context of GMOs, the precautionary principle is seen as an improvement on its predecessors from an ethical perspective. It takes account of a wider range of concepts and recognises a level of common sense concern alongside what can merely be calculated. It would be less valid, however, if it were to deprive people of products which offered benefits and which presented no significant risks. For those with fundamental ethical objections, there is no dilemma. The release of GMOs should never be contemplated. Those who hold more extreme opinions of this nature also tend to be more extreme in their practical approach. As discussed in Chapter 10, some are willing to expend considerable efforts, to break the law and even to put themselves physically at risk in order to oppose developments which do not directly concern them. Rather than a NIMBY (not in my backyard) approach, this is NIABY (not in anybody's backyard). However, lest it should be thought that fundamental values were only the preserve of objectors, it is equally important to point out that such attitudes exist on the other side also. For those for whom nothing should be a barrier to progress and economic growth, the only risks that ultimately matter are the ones which hold back those ideals, and there is no other reason for precaution.

For less extreme views, the dilemma posed by GMOs is that, if developments are to be allowed at all in this area, a decision has to be made at some point to do a trial release, but before society can be certain about either the benefits or the risks. As we have seen, the benefits in a given case can be complex, for example the release of a GMO whose purpose is to reduce the amount of pesticides used in the environment. This is a practical embodiment of the first of the two ethical questions. Above what level of potential risk should we retain a precautionary approach, given that it has its drawbacks in so far as it can inhibit new developments and be a commercial disincentive? What additional benefits or additional knowledge would justify a decision to take such risks, having hitherto taken a precautionary stance?

Societal Legitimation

These questions need to be considered and, given variability in social attitudes and preferences, decisions need to be made on behalf of society which will discriminate among these attitudes and preferences. An ethical commission operating with full public disclosure of its evidence and deliberations would be one way to achieve this and it might help to defuse at least some of the more contentious issues surrounding the release of GMOs.

The intention of including public interest representatives in ethical decision making bodies is that the decisions of the committee should reflect a range of values as well as technical knowledge. It has been found in the past, however, that members who lack scientific expertise have exerted very little influence on committee decisions. As a result, a semblance of consultation has been conveyed, but without having any real substance.[174] Most risk regulators have been educated in the scientific-rationalist tradition and see no place for personal values in decision making on risk, while being unaware of the personal values they themselves bring. When regulators are asked questions about value judgements, the answer generally comes back in a form which has interpreted the words value judgement to mean scientific uncertainty, rather than an association with any system of human values. The intended legitimation of the GMO thus runs the risk of being at best rather hollow, and at worst portrayed as a sleight of hand. Much depends on the extent to which the body is seen to be saying no. If it never does so, people will tend not to believe that its members are exercising their critical faculties, because they see no tangible evidence.

Commercial Pressures

Developing a new chemical pesticide is a very expensive and time consuming business. There is the need to screen tens of thousands of chemicals to find one that is sufficiently active to be useful. The regulatory requirements of safety, quality and efficacy must also be satisfied. When account is taken of all the chemicals that fail tests at the various hurdles, it can cost £20–30 million and take over ten years to develop a new chemical pesticide. To recoup the cost of this enormous investment in research and development, a new pesticide has to be suitable for use on a wide range of crops and pests, in most of the major agricultural systems. This is why the chemical pesticides produced by the agrichemical industry have a wide spectrum of action and are likely to affect species which they were not designed to target.

When the possibility of producing genetically modified biopesticides was first considered, it was generally assumed that they would have a narrower spectrum of action and would be safer and cheaper to produce. Biotechnology was seen as a way of weaning intensive agricultural systems away from the chemical pesticide treadmill. In this context, the research and development history of NP virus from Case Study 5 is informative. Despite the fact that the modified virus was not being developed by a commercial company, successive modifications pushed it inexorably in the direction of increasing the attractiveness of the product to potential customers. There were pressures to widen the spectrum of pests which would be susceptible to it, to increase its effectiveness against the targeted species, and to speed up its action. These all negated its original perceived advantage that it would have a more restricted spectrum of action and therefore be less likely to cause environmental damage through impacts on non-targeted species.

Some of the early modifications were intended to disable the NP virus so that it did not persist in the environment for long periods of time. However, commercial pressures led the developers of a given GMO to increase the length of time for which it remains viable in the environment, in order to give greater assurance of effective pest control to farmers. If applications have to be repeated at frequent intervals to ensure that pest infestations do not recur, a GMO product would not be able to compete in cost and convenience with chemical pesticides. However, from the point of view of the developer of a GMO, a viable market also depends on farmers having to place repeat orders at regular intervals. There is thus a delicate balance commercially as well as environmentally between the persistence and decay of a biopesticide. Such a balance is difficult to achieve. The virus which had been modified to express the scorpion toxin gene appeared less able than the wild-type to sustain the infection in the target species beyond the first generation of the virus. This would have been a problem for its commercial success.

In practice, therefore, what is claimed as an environmental advantage in scientific terms may be eroded significantly when commercial considerations come into play. Indiscriminacy is perceived as being cheaper. The Old Testament biblical injunctions not to harvest right to the edge of a field and to leave a field fallow one year in seven were a reflection not only of the needs of the poor and sound land management, but also that ethical principle that going for maximum productivity has important costs both socially and environmentally.

Global Environmental Ethics

The concerns expressed by GMO risk regulation may be focused mainly on environmental damage within national boundaries, but there is also a global element. GMOs which have the ability to survive and reproduce in the natural environment will also cross national boundaries, either by accidental means or by deliberate policy. Unfortunately, global concerns are not always shared globally. For example, China has major investments in biotechnology. It has already made hundreds of releases of GMOs with little regulation in place, on a scale far greater than the relatively small releases made hitherto in Europe, which are the subject of so much regulation and debate. According to their cultural situation, feeding an enormous population may carry a far greater weight than the effects on the environment. Moreover, there are fears that companies who regard regulatory regimes in industrialised countries as overly restrictive will merely abuse their global responsibility and take advantage of developing countries by testing their GMO products in their territory, when they are not in a position to have a proper system in place.

LESSONS FROM THE UK REGULATORY SYSTEM

The UK government's environmental strategy gave a commitment to taking precautionary action where justified, particularly for new products and processes.[175] When it came to regulating the release of GMOs, a precautionary approach was seen to be justified. This was partly to safeguard the environment, but also to allay possible public fears.

Background Reports and Conflicting Views

The development of UK legislation, in the Environmental Protection Act (EPA) 1990 Part VI, was prompted most immediately by the publication of an EU Directive on regulation of the release of GMOs[176] which also embodied a precautionary approach.[177] The Department of the Environment (now the Department of Environment, Transport and the Regions (DETR)) issued a consultation document beforehand.[178] This proposed a system similar to the step-by-step and case-by-case approach which was already in operation with industry on an informal basis. This was published almost simultaneously with the RCEP report,[179] and there were interesting differences of emphasis between the two. The RCEP recommended more stringent control measures and a greater degree of public access to information than did the DETR.

Voluntary bodies were concerned about the adequacy of the proposals in the DETR consultation paper, and compared them unfavourably with those of the RCEP. Industry lobbyists, on the other hand, felt that the DETR proposals were more realistic and balanced, while those of the RCEP were unnecessarily complex and wide-ranging. However, despite the less stringent recommendations of the DoE consultation document, some industry representatives were already beginning to be uncomfortable with the

precautionary basis of the regulations proposed. They resented the implication that the risks of GMOs are exceptional. They disliked the implication
that GMOs are a form of pollution. They were apprehensive about the
expanding role of the DETR in the regulation of agricultural products. This
was the first wide expression of disquiet on the part of industry, and it was
to build up in the following years into a major campaign for change and
relaxation in the European regulatory system.

The Parliamentary discussion included the highly precautionary
nature of the definition of environmental damage included in the Bill.
Concerns were focused on the duty of care placed upon releasers to
prevent any risk of environmental damage, their liability for environmental damage, and the extent to which *hypothetical* risks could be balanced
against the cost of avoiding the risk and the potential benefits of the
products concerned. The debates also raised wider ethical issues. There
were demands for the setting up of an ethical commission, with proposals
for the remit of this commission ranging from purely ethical concerns to
the assessment of products in terms of their environmental benefit and/or
socio-economic need. For example Brian Gould MP argued:

> *We can easily imagine the impact that GMOs might have in*
> *the hands of self-centred, multinational companies with*
> *laser beam objectives..... It calls into play the increasingly*
> *familiar worries about the moral propriety of mankind*
> *playing God ... some people will be able to manipulate life*
> *for their own specific purposes.*[180]

The Government rejected the idea of an ethical commission partly
because the idea of social progress was equated in their minds with
technological advance, which they felt would be obstructed by such a
commission. The short-sightedness of this view is one of the primary
conclusions of our present study.

Reflection on the Impact of the Act

Part VI of the eventual act of parliament is termed an enabling legislation.
This is a device widely used in UK law which sets out a basic framework
which enables the Government to place more detailed regulations before
parliament at a later date. It provides greater flexibility in allowing for
modifications in response to changing circumstances. Such an approach
has also been criticised, however, because it restricts discussion in parliament, because detailed provisions of the legislation are only set down

when the regulations are drafted. These receive a much less painstaking scrutiny by parliament, and can only be accepted or rejected in their entirety. The system is thus seen to restrict open discussion on changes required from later technological developments, and tends to limit inputs from those outside the civil service to establishment figures.

This latter criticism is especially important with regard to public reassurance, which was intended as one the main aims of the legislative system. In giving a statutory legitimacy to the previous voluntary system of regulating GMO releases, the act could be said to grant a 'ritual normalisation' of organisms which would otherwise be perceived as abnormal.[181] By providing for open scrutiny of all new releases by a panel of relatively independent experts, the GMOs are seen to be transformed from unknown and potentially threatening entities into potentially useful industrial products. Not everyone is convinced.

Some industry groups and scientists felt that the extensive precautions that had been set up would be more likely to alarm the public rather than reassuring them. Viewed from another perspective, the legislation makes possible a greater degree of risk-taking than industry might otherwise be willing to undertake under circumstances of strict liability. The government rejected the recommendation of the RCEP for the imposition of strict liability, where the releaser is liable for damage caused regardless of whether the regulations have been complied with. This provided in effect a state-guaranteed insurance against the potential costs of risk taking to industry.[182] The UK courts have accepted the argument that regulatory compliance limits industry's liability for damage caused, for example, by lead in petrol.[183] The act thus offers anyone who releases a GMO protection from liability, provided they can prove that they have complied with the regulations, while requiring them to pay for environmental reparation if they cannot. The overall effect is to place a large onus on the veracity of the risk assessment and the validity of the means used to do it, and thus on the work of independent experts.

The Role of the Advisory Committee on Releases to the Environment

An important outcome of the Environmental Protection Act was the setting up of the Advisory Committee on Releases to the Environment (ACRE). ACRE was given a combined remit, which included responsibility for the introduction of non-indigenous organisms into the environment, in addition to GMOs. This reinforced the apparent similarity between GMOs and non-indigenous organisms, with consequences for the extent of the perceived risk of GMOs described above. It took over the functions of two existing committees in different departments. To take account of historical associations, the government decided that ACRE should be located in the Health and Safety Executive, financed by the DETR and handled by a joint secretariat involving both departments.[184]

The representation and range of expertise involved in the new committee was the subject of debate. Political appointments to the committee were opposed, and there was general agreement, at least in official circles, that representatives of environmental groups should not play an expert, scientific role in the risk assessment process. One of the precursor committees included a larger number of ecologists, who, as observed above, generally tend to see GMOs as being potentially more risky than the microbiologists and laboratory scientists, who were the main scientific expertise on the other committee. The number of ecologists was increased. The industry lobby was keen to continue representation from trades unions and the Confederation of British Industry (CBI) from some of the companies working in this area. Trades Union Congress (TUC) nominees were widely, if somewhat questionably, regarded as representing the public interest. In resolving these conflicting pressures, CBI and TUC nominees were retained, and the categories of farmer and environmentalist were added to incorporate wider public concerns into ACRE's discussions.

The UK government's intention, in setting up ACRE and devising its remit, was to combine both technical and political judgements in one committee, and to provide reassurance for the public about the new technology at the same time as ensuring its actual safety. A science journalist described ACRE's chairman, John Beringer, as 'clerk-of-works to the New Creation. Like Noah, his committee stands at the gangplank of the ark, punching out tickets to the future'.[185] However, even before ACRE had been properly constituted, industry had begun to lobby strongly, at the UK and European levels, for a relaxation of the strictly precautionary approach to risk regulation.

PRESSURES TO RELAX THE PRECAUTIONARY REGIME

International Differences in Regulatory Regimes

A perceived difference of approach to GMO regulation also began to emerge between the US and Europe, the US system being apparently less precautionary than the European. In the US, GMOs are regulated by the same committees which have been set up to regulate foods, drugs and pesticides and which operate a reactive approach to risk regulation.

Industry representatives began to oppose strongly the European system and accused the European Commission of political hostility towards biotechnology.[186] They set up a lobby group in Brussels, the Senior Advisory Group on Biotechnology (SAGB), funded mainly by multinational companies. SAGB claimed that a precautionary approach would be more time consuming and expensive for industry. It published figures purporting to show that, largely as a result of oppressive regulation, new investment in biotechnology in the EU had dried up.[187] The Environment Directorate (DGXI) of the Commission challenged this view, claiming that the two systems differed only in administrative terms, and that the result for the industry was the same.

Similar pressures were exerted in the UK, where industry alleged that the regulatory system governing the release of GMOs was placing the UK at a competitive disadvantage compared with non-EU countries like the US and Japan. Even within Europe, there are significant differences in national perceptions of risk and the level of implementation allowed within the European Deliberate Release Directive.[188] In Germany, there have been serious public protests and even acts of sabotage directed against the release of GMOs, while in France there has so far been less public concern. Germany is widely regarded as having one of the most restrictive European regulatory systems and has had relatively few releases, whereas France's regime is more relaxed, allowing many more releases.

House of Lords Enquiry and its Effects

The House of Lords Select Committee on Science and Technology set up an enquiry in 1993 to look into the competitiveness of the biotechnology industry.[189] Their report considered that the level of investment and intellectual property rights were probably at least as important as the regulatory system in inhibiting biotechnological development. However, it accepted the view expressed by industry that the regulatory regime was unscientific, because it failed to discriminate between activities which involve real risks and those which do not. The regulatory system was, they felt, too bureaucratic, time consuming and costly and was an unnecessary burden to academic researchers and to industry. They concluded that any regulation which reduces competitiveness should be reviewed critically, particularly when it cannot be justified on scientific or public interest grounds. Without explicitly saying so, this amounted to a rejection of the precautionary approach to risk regulation for GMOs, on the basis of its lack of scientific underpinning, and the economic criterion of its effects on competition.

The report influenced EC regulation, which simplified the applications procedure[190] and reduced the information required to release or market genetically modified higher plants,[191] and also led to the relaxation of the UK Regulations.[192] While the precautionary approach remains in place for many of the GMOs being processed through the regulatory system, regulators have concluded that there is now enough experience with some GMOs, such as genetically modified crops, to be reassured about their safety. Some in industry feel that the changes to the regulatory system have not gone far enough in taking account of the recommendations of the Select Committee. Others are nervous that relaxing the rules too rapidly could alienate public opinion and, in the end, make it more difficult to introduce the products of the new technology to the market place. Environmental groups have, indeed, viewed these developments with great concern.

IS IT WORTH THE RISK?

Aside from the two extremes referred to above, it is impossible to make an overall conclusion from the totality of genetically modified organisms and potential releases. From a point of view of Christian ethics, a number of key values that might lead to decisions one way or the other can be derived from the principles of the precautionary approach. Given that there are conceivable problems, that genetics is still a relatively young science and that many fundamental uncertainties in the basic science remain, it would not be responsible to reject precaution. A trend to relax some areas of the regulatory approach may well be appropriate, but many areas remain where this could not be justified.

The grounds on which such risks may be taken need special scrutiny. It is unethical knowingly to subject others to risks in order to increase the benefits to oneself. In general, the more the primary benefit is economic profit for a few, or a relatively trivial consumer choice for those who can afford it, the less Christian ethics would justify the risk. To return to the pesticide example, many consumers feel that the risks posed by small residues of pesticides, chemical or genetic, on food outweigh the benefits of cheaper, blemish free produce, and are willing to pay a price premium to buy organically produced food. But some simply could not afford the choice. In so far as there is a risk, they may not be able to avoid it.

Underlying this apparently straightforward question of consumer choice there is, however, a complex web of risk related issues faced by farmers throughout the world. They face threats to their crops from pests, weeds and diseases which could, at the extreme, force them out of business. So long as the majority of consumers continue to demand a level of freedom from blemishes in their food that can only be achieved by using pesticides, the organic option will not be open to most farmers. The same applies to food destined for processing. Crops which fail to meet the high quality standards set by these markets will be rejected – a financial risk for the farmer. In heavily populated countries like China and India, questions of survival and the use of chemical inputs to increase yields may be more important than the cosmetic quality of food.

The people who face the highest health risks from pesticides are the farmers who apply them to their crops and they are also the frontline customers for the products of the biotechnology industry. For them, the choice is between continuing to grow crops using pesticides with a known potential to damage health and the environment and using genetically modified crops requiring fewer applications of pesticide but with an unknowable potential to cause environmental harm. From the farmer's perspective, both rationality and ethics may seem to be served by a switch to genetically modified crops.

Perhaps the greatest concern from GMO releases is that somewhere, sometime, there will be a spectacular exception to the generally encouraging picture to date of the effect of GMOs in the environment. Most novel technology eventually has such an unforeseen fallout. Ethically, it seems reasonable to restrict the scale of use, to avoid the more trivial applications and to focus on those where there is a strong element of human or ecological benefit. As demonstrated by the scorpion toxin case study, research which begins with a sound environmental motivation can be overtaken by the commercial element, and in the end the environmental benefits may be overtaken by greater environmental risks.

Risk perception depends a lot on what has happened in living memory. Genetic engineering has the dilemma of emerging at a period in history when European society is much more aware of the drawbacks of new technology than it was a generation before. If the advances of the 1980s had occurred in the early 1950s, few of the constraints and concerns this chapter has debated would have seemed so relevant. And possibly we might now be living with the consequences of both what went right and what went wrong.

8 PATENTING LIFE

INTRODUCTION

Life is special, humans even more so, but biological machines are still machines that now can be altered, cloned and patented. New York Times

The patenting of animals brings up the central ethical issue of reverence for life. Will future generations follow the ethic of this patent policy and view life as mere chemical manufacture and invention with no greater value or meaning than industrial products? Or will a reverence for life ethic prevail over the temptation to turn God-centred life into reduced objects of commerce? Mark Hatfield[193]

Of all the questions to emerge over biotechnology, none has proved more controversial than patenting. Once scientists developed the skills to identify genes and manipulate them, to what extent did that allow them to claim these genetic sequences and constructs, and the organisms which they embodied, as patentable inventions? In what way were such biotechnological advances the same as inventions in the engineering or chemical industries, and in what way different? A heated debate has raged for many years between the biotechnology industry and patent lawyers, on the one hand, and a wide variety of objectors, on the other. The debate first came to prominence over genetically modified animals, when the US Patent Office granted a patent in 1988 on the Harvard oncomouse, described in Case Study 10. In Europe, much of the controversy has centred on the European Union Directive, but the questions it raises go far beyond the legal technicalities of intellectual property, ranging from the role of commerce as a driver of biotechnology to the very nature of living things and human exploitation.

This chapter first explains what a patent is, what qualifies an invention as patentable, and some of the practicalities involved. It then traces the key developments and implications as the scope of patenting began to be widened to include biological systems and animate matter. Some features of the EU Directive are given as an example. With this background, the ethical issues are discussed, first examining the role of ethics in patenting, and then the arguments for and against applying patents to genetic engineering. Finally, some shortcomings are identified not only of the present system of patenting but also the use of ethical appraisal in biotechnology.

WHAT IS PATENTING?

Patenting is a system which provides an inventor with a period of protection from others who might take unfair commercial advantage of their invention. In return the inventor makes a full disclosure of the invention, in the form of a published patent. Patenting is intended to discourage trade secrets and to encourage open disclosure of technical information while protecting the claimed invention.

Much confusion has arisen because of misunderstandings of four basic points. The first is that the patent does not give the owner any right to exploit their invention. It only confers a right to prevent *others* from exploiting it without the permission of the patent holder, and the payment of appropriate royalties. It is only a negative right to stop someone else from doing what is claimed in the patent, not a positive right for the inventor to do it. This means, secondly, that a patent is not concerned with the morality, efficacy, safety or social desirability of the invention *itself*, at least strictly speaking. The invention must still satisfy whatever national regulations exist governing the production or use of the invention, for example concerning animal welfare or safety. As will be seen later, however, this becomes more complicated when applied to biological systems, especially how the ethics of the invention is then assessed.

The third point is that a patent is not intended to prevent anyone from using the knowledge of the invention in pure research. Its purpose is to prevent anyone from the unauthorised marketing of the invention. Lastly, the patent also gives protection only for a restricted period of time – usually 20 years from the date of application, provided that the annual renewal fee is paid.

Most countries, though not all, have some form of patenting system. Applications have to be made specifically to that country. To try to obtain the widest protection, it is therefore usual to submit claims in several places at the same time. In the case of Europe, application can be made to the European Patent Office (EPO) in Munich for a patent which then applies to all the signatory countries to the 1978 European Patent Convention. Details of patent law vary somewhat from one country to another, and rather importantly, as will be seen, between Europe and the US. The European Commission is attempting to harmonise discrepancies between patent laws and practices within the EU, and the World Intellectual Property Organisation is seeking to do so among Europe, the US and Japan.

REQUIREMENTS OF THE PATENT SYSTEM

Under the term of the European Patent Convention, in order to be patentable an invention must be:

- novel and involve an inventive step;
- not obvious to a person skilled in the particular art;
- able to be made or used in any kind of industry, including agriculture,
- supported by a description.

In addition, the patent must pass an ethical criterion that it must be deemed not contrary to *ordre publique* or morality, the exact meaning of which has been much debated but has been re-interpreted in the EU Directive.

Novelty

To satisfy this condition, the invention must be something that was not available prior to the date the patent application was filed. A patent examiner, and also anyone wishing to contest the application, will make a comparison between the description that is submitted and all previous publications in that field, to ensure that there really is a novel step in the claimed invention.

There is a crucial difference between European and US approaches as to what amounts to priority. In Europe, rival priority claims are settled on which was the first to file, ie the earliest application date. A corollary of this requirement is that there must be no public disclosure of the invention whatsoever before filing. A description of the invention in any scientific meeting or publication would invalidate the claim. Consequently, it behoves the inventor to be absolutely secretive until the application has been filed. There are cases where filing can be done very quickly, but frequently it may be a considerable time after the invention was made. As a result, the first to file requirement increases the commercial pressures to restrict scientific discussions, and causes major conflict between patent law and the need for early scientific publication. As academic institutions are increasingly being asked to be more commercial, this can create real difficulties.

In the US, however, a year's grace is allowed between invention and patent application, which allows the inventor to file a claim up to one year after first publication. Competing claims are then settled by comparing actual dates of invention. This appears to confer advantages, but it can also result in protracted and acrimonious legal battles between applicants each claiming primacy of the invention, in which scientists' precise entries in laboratory notebooks can become highly contentious pieces of evidence.

The difference between the two approaches is exemplified by a successful and very lucrative patent granted to Stanford University for recombinant DNA techniques in 1973, following publication in a journal and also after press speculations on its possible uses. In contrast, in 1975 the UK Medical Research Council lost all rights to patent their *hybridoma* technology which produced monoclonal antibodies, because the work had been published first. Consequently, for 20 years a worldwide industry now used this technology free of charge.

The EU Directive allows that any biological material isolated from its natural environment and any invention containing biological material or processes can be patented, with the sole exception of biological processes to produce plants or animals, and plant or animal varieties. Hitherto, the general view of patent offices has been that merely to 'pluck something from the earth' for the first time does not make it patentable, but this is changed by the Directive, raising considerable ethical problems.

Non-Obviousness

To satisfy this criterion, the examiner must be convinced that there really has been an inventive step, which would not have been obvious to someone 'skilled in the art'. According to the European guidelines it must 'go beyond the normal progress of technology' and not 'merely follow plainly or logically from the prior art'. The guidelines note that a new idea may soon seem to be 'obvious'. With the hindsight that is available to those contesting an application, it is much easier to show the relationship between the prior art and the new invention than it may have been for the inventor. The examiner has therefore to have in mind the state of the art at the time of the invention, which often requires expert judgement.

Industrial Applicability

In principle, ideas, notions and discoveries are not patentable. The invention must be embodied in some product, process or apparatus which is capable of industrial application. The EU Directive, however, now implies that natural products can be patented provided they have an industrial application. This is seen by many to undermine the basic distinction between discovery and invention. What a patent lawyer means by the term 'discovery' is not what it means in normal English usage but in essence refers to something natural which no one has yet found a use for. Industrial application can nowadays include agriculture and medicine, with certain exceptions. In Europe there is a specific exclusion of 'methods for treatment of the human or animal body by surgery or therapy and diagnostic methods practised on the human or animal body...'.[194] This does not, however, exclude specific substances for use in any of those treatments. Hence a therapeutic drug is in principle patentable, but the therapy itself is not.

Adequate Disclosure

The application must describe the invention so that a person 'skilled in the art' may reproduce the work carried out by the inventor. A claim that is not sufficiently detailed will inevitably fail as it is not possible to add information once an application has been filed. This requirement presents particular difficulties for some areas of biotechnology, where techniques may be relatively unrepeatable and where there may be subtle, unknown differences between populations of animals which are believed to be identical. This highlights one of the fundamental problems of applying a system of patenting designed around replicable mechanical inventions to inherently non-replicable living organisms.

Exclusion on Moral Grounds

The above are all technical criteria which have to be satisfied. In addition, the European Patent Convention and the EU Directive both include a moral requirement for patentability. They require that inventions would not be granted a patent if their publication or exploitation would be contrary to what is termed '*ordre publique* or morality'.[195] This rather vague concept has now become very much more prominent as a basis to challenge biotechnological inventions, as will be discussed below.

The Scope of the Patent

The scope of the protection being sought may vary considerably. It may be a very specific but significant detail of an existing process, such as a new step in making a genetic modification. It might be an entire range of possible applications, of which the particular example is the first to be discovered. Frequently, a patent will contain a series of clauses which set out the protection which is being claimed, starting broadly and getting progressively more specific. Thus a claim which was based on a particular genetic modification of cotton was filed as a claim for protection for *all* types of genetically engineered cotton. The broader the scope of the claim, however, the more difficult it is to justify. In this case the patent examiner did not agree that such a very wide scope was in fact justified by the basic invention.[196] The scope may also be a function of the degree of maturity of the technology. The initial patents granted in a new area of science are apt to be rather broad and all encompassing, but as the technology develops, applications become narrower, and the patents become much more closely defined. Often this happens only after expensive and protracted litigation or challenges to the broader claim.

Where a patent already exists on a genetic application, someone else may find a new use, not envisaged by the original inventor, and file a patent application on this new use. An accommodation is then often found by a system of cross-licensing, giving the second inventor specific protection, while acknowleging that their invention owed something to the first patent holder.

Obtaining a Patent and Challenging It

In Britain, an initial application is filed at the Patent Office to establish priority. Within a year, a full claim must be submitted, at which point a preliminary examination is made by a patent examiner. After a preliminary search through journal references and patent office files, the application will be published together with the result of the search. At this point, objections to the proposed patent can be raised. In theory, anyone can lodge an objection, but in an area governed by complex legal technicalities, the scope for private objection is inevitably limited. A substantive examination of the patent application is then made. This takes account of the search, any objections, and the four requirements listed above, and the patent is granted, or not. If agreement cannot be reached between the examiner and the applicant (or their patent agent) there is a right of appeal to senior officials within the Patent Office and to the courts. Once a patent has been granted there is a further period of nine months in which it can be challenged.

Some Practical Problems with Patenting

Patenting has become one of the cornerstones of industrial development, especially in biotechnology where patented products are seen as the primary return on the very substantial initial research investment. As a result, and partly in response to the claims of objectors, industry and government are apt to make rather sweeping claims for the unalloyed benefits of patenting, to promote the wellbeing of the biotechnology industry and its ability to secure future inward investment. In reality, the picture is not always quite so rosy, and to set the record straight some of the flaws of the patenting system also need stating.

Firstly, patenting is a very protracted, arduous and expensive business. It can divert a lot of time and resources from other research, in order to deal with the complex legal paperwork, record keeping, queries and technical points required by the lawyers who are preparing the patent application. They may often request additional experimental work or expanded data. These might be to confirm what is being claimed, or to justify a widening of the scope to cover as many related areas as possible in order that the patent can generate the maximum income. As a result, a scientist can lose control of the logical progression of the research work to the dictates of legal and commercial priorities.

The eventual cost of a patent can be substantial. This might be £30,000 for legal and other costs of filing and obtaining a basic patent as quickly as possible, even before the subsequent requirements and correspondence with the patent examiner. If to this are added the costs of preparing and negotiating any royalty deals with organisations wishing to use the invention under licence, and any litigation to defend the patent claims, the bill can run into hundreds of thousands of pounds. This may sometimes outweigh the initial research investment, and might deter smaller organisations from going down the patenting route. A recent study by the European Patent Office[197] found that 'small and medium sized enterprises are especially sceptical about patent protection', for various reasons which included the cost involved, and the fact that it does not guarantee commercial exploitation or necessarily deter imitators. This suggests that, in practice, patenting is a form of investment protection which more readily benefits larger organisations. Patenting also adds costs to the end user of the technology or organism, which may put it beyond the reach of poorer buyers, for example in developing countries, or in some cases restrict the availability of medical treatments in the National Health Service.

In the UK, the culture change in research funding policy has put an emphasis on patentable results as a goal likely to be more successful at securing critical funds. To obtain commercial funding, a research programme increasingly requires promises of a return on investment within five years, in the form of filed patents and projected income from royalties or licences. For many scientists this has come as an unwelcome development. There is also the concern that, being only human, promotion and peer review rewards for success in patenting can adversely influence a scientist's judgement.

Even at its best, patenting will never be a perfect system, but the existence of these problems draws out an important point. It is not simply a smooth, objective technical process, but rather a system of intellectual property which works in a particular practical, social and ethical context. As will be seen, the context becomes crucial when it comes to the implications of applying this system to biological material.

THE APPLICATION OF PATENTING TO BIOLOGICAL MATERIAL AND LIVING ORGANISMS

Patenting was developed around the notion of physical inventions and industrial processes. In general, it has not been applied to biological material until recent times. Although Pasteur obtained a US patent in 1873 for 'yeast free from germs of disease as an article of manufacture', the US courts later decided that the 'discovery of some of the handiwork of Nature' was unpatentable. One reason is the fundamental ethical problem of whether it is appropriate to claim invention of a product of nature. More pragmatic problems are that if the invention is alive, it is hard to define it in a patent, and it is also capable of self-propagation. What then is the status of the second and subsequent generations which could be bred by the purchaser? This was a question already faced earlier when the idea of intellectual property protection was applied to the conventional breeding of plants and animals.

Plant Breeders' Rights

In Europe, patent law was considered unsuitable for protecting new plant varieties developed by traditional breeding. European patent law subsequently excluded animal and plant varieties and 'any essentially biological process for the production of animals or plants' and this was confirmed in the EU Directive. In 1960 national Plant Breeder's Rights (Plant Variety Rights) were established, followed in 1961 by the International Union for the Protection of New Varieties of Plant (UPOV Convention). This convention provides the breeder of plants with the legal ownership of the new variety, and the right to multiply and sell seed of a specific variety.

It also upholds the principle known as 'farmers' privilege', by which farmers are allowed to keep some of the seed from one harvest to the next, for further breeding, without needing the consent of the breeder. This attempts to balance the interest of the breeder, in preventing the sale of the seed by a competitor, while allowing widespread use of the improved seed simply for the cost of one purchase of seed. It should be pointed out that this is not always to the farmer's own advantage to breed from the initial purchase of seed, because in certain cases the second generation would lose something of the benefit of the original seed. In 1991, the UPOV Convention was strengthened to include 'essentially derived' varieties, which comes closer to patent protection, and is said no longer to prohibit plant breeders from acquiring either patent or variety rights.

No equivalent system of animal breeder's rights exists, but the EC Directive on patenting would extend farmers' privilege to animals, provided they do not do so for commercial reproduction. It is a vexed question as to whether the patenting or breeders' rights system is the more appropriate. A patent would be broader and cover variants of the core invention as well as the variety itself. As with patent law, research use is allowed. Neither system prohibits the free use of existing, known germplasm.

US Patenting of Living Organisms

Over the last 30 years, the whole picture of intellectual property rights for biological material has changed, beginning in 1971, when an application was made by the General Electric Company and Ananda Chakrabarty for a patent on a genetically engineered microbe which could digest hydrocarbons. This became a landmark case. The US Patent and Trademark Office (PTO) first rejected the application on the grounds that animate life forms were not patentable. However the Court of Customs and Patent Appeals reversed the decision and allowed the application on the basis of their opinion that 'the fact that micro-organisms ... are alive ... (is) without legal significance'. The court tried to limit the significance of its decision by claiming that micro-organisms are patentable because they are 'more akin to inanimate chemical compositions such as reactants, reagents, and catalysts, than they are to horses and honeybees or raspberries and roses'. The PTO challenged this decision and eventually the US Supreme Court gave a narrow decision, by five votes to four, that the oil-eating microbe was not a product of nature and was therefore patentable. The majority judgement stated that 'the relevant distinction was not between living and inanimate things' but whether living products could be seen as 'human made inventions'.[198] This case established, at least in the legal system, that in principle living organisms could be patented, and has subsequently influenced all other industrially developed countries. Micro-organism patents are now routinely granted in the US, Japan and Europe.

In 1985 the US PTO ruled that the patenting of microbes could be extended to include the patenting of genetically engineered plants, seeds and plant tissue. In 1987 they further ruled that the ruling could include all 'multicellular living organisms, including animals'. Humans were excluded on the grounds of antislavery laws, but not embryos, foetuses or genetically engineered human tissue. The extension of patenting to animals, beginning with the granting of the US patent on the Harvard oncomouse in 1988, stirred a widespread outcry. It led to much debate in the US Congress and elsewhere. Despite many attempts, no laws were passed to oppose the decision. As it happens, the Harvard oncomouse patent appears to have rebounded, because it has been reported that few have wanted to take up the licence.[199] The commercial possibilities opened up by the decision, including livestock patenting, are said to have contributed to US biotechnology companies lobbying to dissuade President Bush from signing the biodiversity treaty at the Earth Summit in Rio in 1992.

BOX 8.1 TYPES OF BIOLOGICAL MATERIAL INVOLVED IN PATENTING

- Specified 'natural' DNA sequences with industrial application;
- Modified DNA sequences or gene constructs;
- Whole micro-organisms with industrial application, such as to create vaccines;
- Whole animals or plants containing genetic modifications (but not animal or plant varieties);
- The processes or techniques to introduce a genetic change or a modified whole organism.

Examples of biotechnological patents which have been awarded are many enzymes used in industrial applications, proteins useful in medical therapies (but not the therapies themselves), and animal cell lines with antibodies against particular antigens.

The European Union Patenting Directive

There has been a more concerted opposition to biotechnology patenting in Europe, centred around the Harvard oncomouse and the EU Directive. After an initial refusal, the European Patent Office (EPO) granted a European patent on the Harvard oncomouse in 1990, but objections were filed by anti-vivisection, animal rights, environmental and church groups, on grounds which included morality and animal suffering. By early 1998 the EPO had still not given its decision. Many European patents have since been filed for other transgenic animals.

In the meantime, focus largely shifted to the EU Directive. This was initially written in 1988, seeking ostensibly to produce a harmonised EU-wide legislation, but more significantly to enshrine in law the principle of allowing the patenting of transgenic animals, plants and micro-organisms, and also sections of the human genome. The EPO was engaged in drawing up the Directive and would adapt its work in accordance with the legislation. However, it became something of a *cause célèbre*, involving a confrontation between industry and environmental interests, and also between the processes in the parliament and the Commission. After protracted negotiations, intense lobbying and numerous amendments, the original Directive was rejected by a substantial majority of the European Parliament in March 1995, to the surprise of most commentators and the great embarrassment of the European Commission. Although claimed as a victory for the Green lobby, it was really more of a stalemate, since the EPO could continue to grant biotechnology patents, in a piecemeal fashion. Under considerable pressure, the EC prepared a new draft Directive in late 1995. Although its wording addressed some concerns, it soon became clear that the EC had not sufficiently taken on board the ethical issues which underlay so much of the opposition to the previous directive.

The preamble to the Directive declares the intention to address the fundamental question of whether patenting can be extended to animate and biological matter. Its answer is that no invention should be refused patent protection for the sole reason that biological material is involved. Its philosophy is thus to declare that for the purposes of patenting there is no essential distinction between the biological and the non-biological, with certain named exceptions. It also blurs something of the distinction between a discovery and an invention by declaring that 'Biological material which is isolated from its natural environment or produced by means of a technical process may be the subject of an invention, even if it previously occurred in nature.' It is the morality of these basic stances that is the root of much of the opposition. The Directive justified its stance primarily on legal and commercial arguments, pointing to the precedence of decisions made in settling patent law cases, but it notably failed to provided an ethical justification to address the underlying objections. As a result the same controversies duly began again.

Its permissive approach of 'yes, provided' would allow for all micro-biological processes and products to be patentable, and all plants and animals, except those produced by processes 'consisting entirely of natural phenomena' , and except for plant and animal varieties. Any plants and animals produced where a *non-biological* step was involved are patentable, which would include any which were genetically modified. Two fundamental ethical points are made, but again no proper basis is provided for either. One is to draw a distinction in status between micro-organisms and all other types of living organism. The other is that human intervention in an animal or plant can in principle render that animal or plant a patentable invention, even by a change of a single gene.

There are highly controversial articles relating to genetic material of human origin, which lie beyond the scope of this book. The Directive also specifies a number of exceptions which would fall under the category of 'contrary to *ordre publique* or morality'. Again, these are chiefly in certain human applications, but they also include an invention involving suffering to animals 'without any substantial medical benefit to man or animal'. This reflects an acknowledgement of wider ethical concerns represented, for example, by a significant decision by the EPO to refuse a patent for a mouse genetically engineered to increase hair production, used for research into baldness,[200] on the grounds that the animal was suffering for trivial purposes. It has been argued that even where there was substantial medical benefit there needs also to be an element assessing the proportionality of suffering, but unfortunately this was rejected by the European Parliament. Others have argued that such detailed exclusions are inappropriate to the sphere of patenting, and simply represent a political compromise to help the measure through the European Parliament.

The passing of the Directive by the European Parliament in May 1998 is likely to be something of a watershed. The European Commission and the industry argued that a strong directive is needed to maintain a European biotechnology industry capable of holding its own against competition from the US and Japan where much more permissive patent regimes operate. Eventually this view prevailed. To environmentalists, many religious groups and others, including many Members of the European Parliament (MEPs), a Directive has been passed embodying fundamental ethical problems, not only in its wording but also in the inadequate way in which the ethical element itself has been tackled.

PATENTING AND ETHICS

How Does Patenting Relate to Ethical Judgements?

Having considered the nature of patenting, the complex regulatory structure and some of its practicalities, we now turn to its ethical implications. It is often argued that moral issues lie outwith the scope of patenting. Patenting should simply be a matter of technical assessment against strictly limited criteria, such as novelty and utility, and should not involve ethical judgements.[201] A patent is a legal means to protect the inventor from having someone else use it, and confers nothing about the ethical acceptability of the invention. It is also said that a patent office is not the proper place to be making such judgements. Many patent lawyers do not wish to make decisions on the basis of morality or desirability, considering it outside their competence. At first sight there is some force in these arguments, but on reflection there are several problems also.[202]

The view that ethics lies outwith patenting amounts to a statement of principle in itself. It is an essentially reductionist view. Patenting is seen as no more than a technical process which can be treated in isolation from normal life. This is unrealistic. From a Christian standpoint, nothing, no matter how technical, lies outwith God's rule and hence beyond the scope of moral and ethical judgement. Indeed, as will be seen later in this chapter, the very concept of patenting already embodies a particular set of moral judgements about property, human invention and justice, which express certain societal values. This does not mean that there are no reasonable criteria for making technical assessments, or that the judgements are invalid, but all such criteria and the judgements made from them are inevitably value laden.

Secondly, the examiner's assessment of a patent claim is not as purely objective as it might seem. Take the criterion that an invention should not be 'obvious to one practised in the art', for example. It is quite common that different research groups make very similar discoveries quite independently. If they file rival claims, it may require a very fine judgement on the part of the patent examiner to decide whether the invention was obvious or not, especially if it involved a considerable intuitive leap. This is not simply an objective technical procedure, but an expert value judgement.

Thirdly, to suggest that patenting does not involve moral judgement is a *non sequitur* in that European patent law explicitly admits the possibility of exclusion on the grounds that the use or manufacture of the invention would be contrary to 'ordre publique' and morality. This is unquestionably an ethical judgement. It would exclude a letter bomb since there was no legitimate social reason for its use, whereas guns and other weapons can be patented. Formerly this criterion was seldom cited, but as soon as material relating to animate matter is introduced into the sphere of patenting, then *de facto* the ethical dimension becomes a much more important element. This view is clearly stated in the 1996 opinion of the EC Group of Advisors on Bioethics in relation to genetic material of human origin:

> *The consideration of patentability criteria resulting from the usual technical requirements of novelty, inventive step and industrial application, must also take into account consideration of these ethical principles.*

This has been further reinforced in the EU Directive, which has added a set of exclusion clauses for certain biotechnology applications like human cloning and germline gene therapy. The list is, however, somewhat of an arbitrary compilation of issues known to be politically sensitive to the European Parliament, rather than the outcome of a rigorous consideration of the full range of relevant ethical issues.

The claim that a patent confers no judgement about the ethical acceptability of the invention concerned is thus an exaggeration. Moreover, there is also the social dimension. It has been argued that patenting is not an isolated technical exercise, but is embedded in societal values. It also depends on society's perceptions. In the public eye the granting of a particular patent may be regarded as conveying an implicit official value judgement that the thing itself is morally acceptable, since it is of enough value to be worth protecting. If this is how it is perceived, then *de facto* an ethical significance is conferred on patenting. Public perception may not accord with the experts' understanding of what is meant, but, as in the parallel case of risk perception, it is unwise and possibly even irrelevant to discount it as mere ignorance.

The extension of the notion of patenting to living organisms has highlighted the ethical dimension to patenting in a new way, even though it was implicit already. If it were maintained that moral and ethical judgements were indeed outside the scope of a patent assessment, then it might logically be concluded that living organisms, whether modified or not, should lie outside the scope of patenting, since moral questions cannot be avoided. On the other hand, if an ethical dimension is admitted over living organisms, it would be illogical not to consider the ethical implications of some chemical patents also.[203]

Possible Ethical Approaches to Patenting Genetically Modified Organisms

The concern is with patenting and its effects, rather than the ethical concerns about the particular genetic modifications. The fundamental issues are whether it is acceptable or not to patent living things, and the underlying questions implied by it. A decision on whether or not to issue a patent may also have many effects – for the organism concerned, the consumer, scientists, companies, farmers, doctors, potential users or non-users of the invention. There may be an impact on society as a whole. There are also implications for how the laws that regulate the treatment of animals, and the patenting procedure, relate and respond to each other. Several of these ethical issues go to the heart of theological and philosophical judgements about the nature and meaning of life, including questions of ownership, distinctions assumed between particular life forms, and what is our fundamental attitude towards living things which have a utility to humans.[204,205]

ISSUES OF PRINCIPLE CONCERNING PATENTING

Ethical Principles Assumed in Patenting

Patenting already assumes and enshrines a number of fundamental ethical 'goods' which are taken as normative:

- the free and open access of information and its benefits to a society;
- the idea of reward for inventiveness within a competitive commercial culture;
- the principle of justice to protect that reward from abuse.

Scientific research is conducted for a variety of reasons. Amongst other things, it is done to satisfy the interests, curiosity and career of the researcher, to enhance the reputation and future funding prospects of an academic or research institution, and to produce profit for an individual or organisation. Research is also a societal activity, done on behalf and with the blessing of the society which has provided the structures and the means for the research to be done. There is a fundamental principle that knowledge gained from research should therefore be freely available to all, as opposed to remaining the private property of a few. The progress of science depends to a large degree on the sharing and discussion of results amongst the research community of those working in that field, which also extends internationally.

There is also a principle of reward for inventiveness. If someone makes something new or inventive, they are entitled to put it on the market and gain a profit and livelihood from selling it, subject to the normal rules by which a society governs what may or may not be bought or sold. In so far as the inventor relies on the knowledge which he or she has gained to be able to sell the invention, that knowledge is not something which is freely available to all, if the result would then be that someone else would begin to market the same invention under their own name. The particular knowledge which enables the inventor to sell the invention thus itself has an element of property, and to this extent could be said to have a commercial value. This is in tension with the notion that knowledge should be freely available to all.

Patenting in the first place assumes these two ethical norms, and then seeks to provide a way to resolve the tension which arises between them. At the root of the notion of intellectual property in general is a principle of justice. Some measures are needed to restrain a tendency of human nature to take something for nothing if no one bothers to stop them. Patenting is a response to mitigate the abuse of people appropriating the fruits of another's hard and expensive labour, without redress. This is a reflection of a failure of relationship and mutual respect and responsibility, a response of a failing in human nature under commercial pressure. In the competitive commercial world, this aspect of human relations is in effect reduced to a legislative framework.

The primary ethical question is to what extent, if any, the principles underlying patenting still apply when the subject of the patent involves living organisms and biological material. A corollary of this question is whether patenting is in fact the correct way of dealing with technological innovation involving living creatures and life forms, or whether some other approach is called for.

Arguments of Principle in Favour of Patenting Genetically Modified Organisms

In favour of extending patenting to this area, it is argued that genetic engineering, whilst mostly a biological process, involves a crucial step which is not biological. Since human intervention is involved – either by devising inventive methods for the isolation of the organisms or by inducing a modification in the organism – then it is claimed there should be no reason in principle why this inventiveness should not be rewarded by a patent. Hence, if a new organism is produced in a defined and different way it may be possible to obtain a patent either for the organism or the method of production. The biotechnology industry maintains moreover that there should be no discrimination against legal protection of intellectual property in this field compared with other technologies. The EU Directive and US Supreme Court decision are broadly in support of this industry position.

A senior UK patent expert has noted that a change of culture has occurred over how life forms are regarded with respect to patenting.[206]

> *Historically, the patent system came to birth to meet industrial needs. Industry was perceived as activities carried on inside factories... Manufacture was the key word. Agriculture was felt to be outside the realm of patent law. Living things were also assumed to be excluded as being products of nature rather than products of manufacture... This restricted view no longer persists in most industrialised countries. Thus the European Patent Convention of 1973 declares agriculture to be a kind of industry. Nevertheless vestiges of the old idea can still be found. ... From the point of view of industrial and social policy, the application of technology to living organisms as industrial tools or products should raise no objection in principle.*

It is also argued that to patent does not in itself reduce the thing patented to being nothing but a commercial entity. Thus when applied to living organisms, all the other aspects of the creature or plant remain exactly as they were, whether it is patented or not. All that has been done is to prevent someone else claiming rights over an inventive activity which has been done in relation to the organism and its potential uses. From a biblical standpoint, it is argued, this has not diminished the creatureliness of the creature.

Arguments of Principle Against Patenting Genetically Modified Organisms

Those opposed to the above viewpoint argue that to patent something living marks an unwarranted extrapolation of the principle of patenting to fields for which it is either legally inappropriate or morally objectionable. Implicit in the argument for patenting animate matter is an assumption that living organisms and human genetic material can be regarded as an extension of inanimate objects. It is argued that to claim to have invented something that is alive like a transgenic mouse, as if this were no different from a mechanical invention like a mouse trap, is to violate something basic to the identity of something which has a life in itself. Normally we make a clear distinction between what is alive and what does not have life, but this would cast aside that distinction, and the associations we normally make about it.

Violation of Life as a Gift

The primary problem is that no one can claim to have invented something which is alive. To claim invention by changing only two genes out of an animal's entire genetic complement is a gross exaggeration. The basic animal remains a biological product of the process of sexual reproduction. If it has an 'inventor', it is God. There may indeed have been inventiveness in the gene construct and its use in a particular animal. It could be argued that the construct or the use could be patented, but not the modified animal itself. To extend the scope of the patent to include the entire animal incorporating the construct would be an unwarranted claim over the nature of a living organism. This is a failure of due respect of the organism for what it is, disregarding or belittling the distinction of life and non-life. Some would go further and consider the approach to nature implied in this as amounting to a kind of blasphemy in offending against the creator and against the creation as gift of the creator. To patent living things could be said to challenge the Christian concepts of life as a free gift of God, with the sense of gratitude which this brings, and of life as a shared inheritance, which are also found in other religious faiths. The concept of gift is seen as fundamentally incompatible with the idea of human invention of a living organism itself.

The sense of gratitude for a gift can also be obscured by the commercial context, if the main emphasis becomes that of seeing creatures primarily as means to profit. Respect and reverence for life are essential within religious perspectives and also in most secular ones. This idea of reverence is an essential corrective to some modern views of life and nature which see fellow creatures as very little more than production machines or resource banks at the disposal of humans. This was expressed, significantly, by a group of American religious leaders who protested at the patenting of animals when the decision was first taken in 1987:

> *The decision of the US Patent Office to allow the patenting of genetically engineered animals presents fundamental dangers to humanity's relationship with the natural world.*

Reverence for all life created by God may be eroded by subtle economic pressures to view animal life as if it were an industrial product invented and manufactured by humans.[207]

An Unwarranted Extension of the Principle of Ownership

Some argue that patenting represents an extension of the concept of ownership which is questionable ethically and theologically. It marks a significant further step in a larger process of the commodification of life. To claim ownership of something is arguably a lesser and more instrumental claim than the ontological claim of having invented it. Consequently the objection on ownership grounds is less important ethically than the violation of life argument, but it raises some important points, nonetheless.

Those in favour of patenting in this area argue that farmers own their livestock or crops, and people own pets, so there is nothing either new or undesirable about the idea. If then we accept that farm livestock can be used to make a profit, why should there be a special concern about patenting it?[208] It is important, however, to clarify what is meant by ownership. The word covers many different senses of property relations. A washing machine we have bought, a poem we have written, a flat we have rented, a piece of land for which we are a trustee, and computer software for which we have bought a licence, all represent different concepts which relate to ownership. Few would object to the idea of a farmer having certain property rights over a dairy cow, but to say that someone owns something that has a life of its own, like a plant, or especially a pet or farm animal which can get up and walk away, is clearly not the same as owning an inanimate object which cannot.

Moreover, it is generally accepted in society that not everything is or ought to be a commodity. Distinctions are made and lines drawn. This raises the question of why certain lines are drawn in particular places. The Bioethics Convention of the Council of Europe, for example, lays down that the human body and its parts should not as such be a source of profit.[209] Since the nineteenth century, we have recoiled from the idea of owning slaves, as fundamentally contrary to the dignity and nature of a human being. Each human being has a life of their own. It is not for another to have the right to control, manipulate, purchase, sell or destroy a human life, which ownership would imply. Just as we draw a distinction between what is human and what is not, there should be some distinction in ownership of living things, which recognises a different sort of owning in ownership of a cow from owning a washing machine or a painting.

What then does this tell us about the ownership of intellectual property, like the results of a research and development investment, be it living or inorganic? We make distinctions between ownership regarding humans, animals and plants, and mechanical tools. Consequently, it is argued that intellectual property needs some means of making a distinction for what are primarily products of nature, albeit with small changes brought about by human agency and inventiveness. Since, historically, patents are associated with products of industrial manufacture, to allow a patent on something living implies that the living thing has no essential difference from an industrial commodity to be bought and sold.

Some maintain that the above analogies based on ownership are irrelevant, since patenting is strictly only a negative right – to stop another acting. Because it does not give the right for the holder to manufacture, it is not a right of ownership of the thing. This seems to be splitting hairs, since even a negative right implies that the holder must first have a claim over the invention as their own and not another person's. Some also seek to point out that in patenting it is only the idea over which property is claimed, and not, say, the living organism itself. The idea could not be patented, however, unless it could be embodied in an actual invention or process with industrial utility, and so it is perhaps invidious to say that what is protected is nothing but an idea. Moreover, if patents can be refused on the grounds that the invention, if made, would be contrary to public order, it is clear that something more than an idea is involved.

Is a Genetically Engineered Organism a Product of Nature or Industry?

Following from the above discussion, a key question emerges as to whether animate matter should be regarded as a product of nature and therefore unpatentable, or a product of industry and therefore patentable, or perhaps both. It is in this sense that patenting has become a focus of the wider question of biotechnological manipulation of living creatures. It is significant at this point to recall that patent law used to exclude 'products of nature'. Following the US Supreme Court decision in the Chakrabarty case and the European Patent Convention's definition of agriculture as a kind of industry, this is no longer held to be the case. It becomes important to ask what it is that has changed, and as a result of what pressures or events, and whether the change is valid. According to Crespi's analysis,[210] such a change had already happened by 1989, and though 'vestiges of the old idea' remain, they are interpreted as a 'restricted view' and anachronistic. It is almost a source of surprise that this view has not entirely died out.

What has changed is thus the application of industrial principles and practices derived from inanimate objects, to agriculture and living things generally. The ethical concern over this is that, at least among industrial, economic and political circles, a fundamental paradigm shift appears to have occurred in understanding the nature of living things. Animals, plants and micro-organisms are now potentially products of industrial manufacture, like the proverbial 'widget'. This could be seen as a commodification of life, a reduction of the value of life and nature to the merely economic, and living organisms to the level of industrial tools. This is at the root of much of the opposition to the notion of patenting life forms. The intuition is that this is one further, highly significant step in a process towards regarding living organisms as, in essence, nothing more than biological machines for human utility, with no distinction from inanimate objects.

Put this way, the products of industry argument has a strong ethical complaint to answer. In its defence, however, it is argued that the true paradigm shift is not from nature to industry, but rather away from a restricted view that life forms are *just* products of nature, which ignores their industrial and commercial potential. Now they are seen as products of industry *as well* as of nature. In essence, it is no more than a techno-logical extension of the idea that a farmer can own, breed and rear animals for them to be killed for food. As discussed in Chapter 4, animal welfare is not necessarily at odds with genetic engineering, and many who are involved in it care a great deal for their animals. Similarly, a patented animal would not necessarily be any less well looked after just for having been patented. Unfortunately, it must also be said that there are enough evidences of misuse of technology, with harm towards both creatures and the environment, to indicate how easy it can be for the product of indus-try approach, in effect, to take over. There remains an inherent tension between the two notions, and the opposition to patenting living organ-isms is to a large degree a response to that tension. It is therefore of much concern that the clause in the EU Directive which excluded from patentability inventions that involved disproportionate suffering to animals has been reduced to the much lesser requirement that there simply be a substantial benefit to humans (or animals). This is a trend in the wrong direction in relation to the issues discussed in Chapter 5.

Summary of the Arguments of Principle

The arguments of principle against patenting genetically modified organ-isms are in part fundamental questions of being, but they are also to do with association and reminder. Some of our group maintain that if animate matter is as patentable as inanimate, not only is a fundamental ontological distinction being blurred, but also the wrong signal is being sent. Christian theology is by nature sceptical of human nature and realistic about human habits. Under competitive commercial pressure, the tendency to sail as close as possible to the wind of what is permissible in law leads to the temptation to regard animate matter solely or mainly in an instrumental way, as though little more than a means towards an end. There have been too many cases of abuse to dismiss. There need therefore to be reminders in the intellectual property realm, as well as in other areas, that a wholly anthropocentric perspective of utility to humans is insufficient. This means that living organisms should remain unpatentable in themselves, whether genetically modified or not. Others of the group, however, while recognising some of the problems, do not see that in principle patenting should be restricted in this way. For them, the associations are not of such force as to constitute a prohibition, and the emphasis should be the need for ethical behaviour in the implementation of the invention.

CONSEQUENTIAL ISSUES ABOUT PATENTING AND GENETIC ENGINEERING

If most of the arguments against patenting of genetically modified organisms are ones of principle, most of the reasons given in favour are consequential. In essence, these are the beneficial effects of patenting in general, as applied to the particularities of the case of genetics. These are to protect research investment and encourage further research, while enabling the public dissemination of the knowledge of the research. If the patenting of biotechnological inventions were not allowed, it is argued that two undesirable things would follow – investment in genetic research would cease or move to other countries, and information of benefit to humanity would not be disclosed. These are now discussed and examined, along with some arguments against.

Patenting Provides Protection for Research Investment, Encouraging Further Research

The primary consequential argument in favour of patenting animate matter in general, and genetically modified organisms in particular, is that it is essential in order to encourage strong research investment in the field, and thus to promote new technological, medical and agricultural advances. As will be seen, this is in part a function of the way genetic research is currently set up, and the framework in which it operates.

Genetic engineering research is expensive in itself, and by no means certain in its outcome. Despite the huge advances, many of the techniques of introducing a modification are as yet still rather hit and miss, and not very efficient. If successful, an application in food or medicine will require long and extensive testing, before it can come to market. All this means that it requires significant up front investment, which carries a fairly high financial risk before a viable product is ready, and before the stringent environmental and safety requirements have been satisfied. If that investment is not protected in some way, then there is no protection for a discoverer, and especially for the biotechnology companies which make genetics their business, against poaching by their rivals.

If biotechnological patents were not allowed, it is maintained that companies simply would not invest in genetic research at all. As a result, new medical treatments, improved agricultural and environmental processes, and enhanced ways of feeding the world would be held back, or perhaps never happen at all. On the face of it, this would constitute a very strong ethical argument, but, short of having a worldwide ban on all such patents, the question is largely theoretical. This is because in practice companies would more likely move their research operations to a country where there are fewer restrictions, or sell them abroad if they could not move. It would not be a case of whether the research would be done, but *where* it would be done. Only if pieces of significant research would not be done anywhere is there an ethical case for patenting on the basis of not inhibiting research as such. The real question at issue is thus one of the competitive status of national research in genetics. It is usually cited with reference to the relative competitive position of UK or European biotechnology, and the risk that the research will still be done, but done in other countries which have, or are said to have, freer regulatory regimes, notably in the US and Asia.

For some, the argument ends here, since a healthy future for the UK biotechnology industry is an ethical good of almost supreme importance, given its anticipated economic and employment benefits, as well as its positive contribution to the general self-esteem and wellbeing of the nation. Some care needs to be taken that it is not made into an ethical absolute before which other ethical considerations are dismissed, but there is undoubtedly much force in the argument. What must be examined more carefully, however, is the assumption that patenting is indeed the necessary prerequisite to this ethical goal, and, if so, what exactly would need to be patented. In particular, is it necessary to patent a genetically modified organism itself? A strict answer would involve a technical debate which lies beyond our present scope. If adequate protection can be obtained by patenting the less ethically sensitive aspects, such as the gene construct, or the use of the construct or the organism to make a specified product, then there would be no good ethical reason to patent the organism itself, on the grounds that it was necessary to promote national research.

Pragmatically, a broad patent covering a genetically modified organism itself is the simplest option for the convenience of a company. It is commonly said to be easier to uphold and defend a product patent than a process patent, and much of the drive towards patenting the organism in itself probably stems from this. Given the existence of other ethical criteria against patenting the organism itself, convenience would be an insufficient reason ethically to do so. To be sufficient, there would need to be a case that the protection would be so inadequate that the research would not be done otherwise.

A further question arises over what exactly a company wishes to protect. This is highlighted by a claim in a public relations brochure of a pharmaceuticals company in relation to the human genome:

> *While genes themselves are essential to life, the intellectual effort required to find them, decipher their DNA structures, understand the roles they play, and identify their use in curing disease – elevate them beyond the status of mere discoveries. In this sense, DNA molecules are inventions that can legitimately be patented.*[211]

On the face of it, this remarkable statement defies common sense, by implying that a discovery becomes transformed into an invention simply if it requires a certain level of intellectual effort to find out what it is. The bottom line is that the investor wishes to protect the investment it put into making the discovery of a natural product, simply because it cost a lot of money to do so. This has left the realms of the legitimate protection of an invention, and is more to do with insuring against lost research investment. Ethically this is especially disturbing. The EU Directive has already been strongly challenged on ethical grounds that it has eroded what would *normally* be understood as the distinction between a discovery in nature and a human invention. The claim of the company would remove the notion of discovery altogether, because the discoverer could always claim that intellectual effort was required. For many people, this would be letting commercial demands go much too far.

There is a more general concern that the benefits of patenting are biased toward larger companies who can afford to undergo litigation to challenge contraventions of their patents.[212] A small organisation may believe its patent rights are being abused by a larger one, but it may not have the resources to risk attempting to prove it in the courts, against legal expertise which the larger organisation can call upon.

Once a climate of patenting has begun in a particular field, it tends to have a knock-on effect on policy in other countries. As a result, any organisation or country with an eye on its markets abroad can feel almost compelled to patent, regardless of any other considerations. At one point patents were being filed in the US for sections of the human genome which had been identified but which had no immediate utility. This produced a flurry of defensive applications elsewhere. The rush to patent everything in sight eventually got sufficiently out of hand that a moratorium was called, albeit only temporary. There is always a danger that the ethic of competition may simply degrade into the law of the jungle, which in an ethically sensitive area like genetic engineering is a very poor guide for public policy.

Opponents of patenting argue that the assertion of discouraging research is as much a tactical defence mechanism to stave off any threat to the existing system, than a substantial argument. No one knows for certain what would happen if patenting of the products of recombinant DNA research were ever disallowed globally. There is at least a case to be answered that if there was a good financial return in prospect, a company would still do the research, but keep the information strictly in house. Moreover, the portrayal of patenting as if it were an unmitigated boon and the saviour of the biotechnology industry falls somewhat short of the reality of the matter. As has been discussed above, patenting is almost invariably expensive, and it can also prove diversionary and litigious for applicants, holders and challengers alike. Again, faced with this situation, some firms have decided to keep their work secret. In such cases the ethical question shifts to one of disclosure rather than discouraging invention as such.

Patenting Encourages the Disclosure and Sharing of Research Information

By requiring the disclosure of information pertinent to the patent application, the patenting process ensures that information is available to everyone, rather than being kept as a trade secret. If patents were not allowed either on, or regarding, genetically modified organisms, the information would be kept in house as trade secrets. Certain advances in knowledge which a company makes, which could be of general benefit to humankind, would thus remain locked away as the private property of certain corporations. Any beneficial applications of that knowledge would, moreover, be restricted to those which that company decides to invest in. The free flow of scientific information would be impeded, and science itself, in those areas, would be held back, as would medical and agricultural benefits.

Stated in this way, this is a very strong argument in favour of patenting in this area. The key ethical question is then: 'Is it necessary to patent living organisms as such to achieve this end of open disclosure of information?' Much the same logic applies as was discussed above. There are also some practical questions which need to be considered, as to how effective patenting really is in encouraging disclosure.

The European and UK system absolutely forbids public disclosure of research results until the patent is filed. This can put considerable pressure on researchers to keep quiet about results they would normally have published at the first opportunity. This is especially the case for academic and research institute work, where the normal culture is one of open publication. By comparison, a patenting culture will mean that the information is made available more slowly, and runs the risk of being beaten to the post if the field is a very competitive one. If it turns out that the results look unlikely to lead to a viable patent, much time will probably have been lost before the information can be made known. If it is viable, it may mean a further high risk delay time to collect enough additional data to satisfy the patent lawyers. If competitor groups publish in the public domain in the meantime, the original inventor would then be robbed both of the intellectual credit for having made the discovery first, and of the potential UK or European patent because of 'published prior art'. Thus in some cases the secrecy and delays implicit in this approach may reduce the rate of scientific progress. Situations also occur in which different groups unknowingly repeat the work already done by others. Similarly, the inability to communicate prevents the exchange of ideas until much later than would have otherwise been the case.

Arguably, the US patent system is more effective in this respect. It allows a period of grace of one year after the publication of the patent application, to allow for claims of precedence. This allows for prior publication of the invention in the scientific literature without compromising the possible patent. The disadvantage is that this can, and often does, lead to bitter litigation, as rival companies file claims for having been first to the invention. The exact time and date of records in laboratory notebooks can become the subject of courtroom controversy.

Generally, but not automatically, the subject of the patent should be made available for research purposes to a third party. Thus patenting is not supposed to impede the research community at large. In practice, this necessitates signing a fairly restrictive agreement giving the patent holder access to any further results and applications of the invention, which may discourage the researcher from applying.

Strategic patenting is in some cases used to close down areas of research which the patent holder has no intention of pursuing, but does not want competitors to work on either. When this is allowed to occur, patenting is a discouragement to open sharing of research. Some companies may patent a beneficial technology in order to prevent its further use and development, if it would be more profitable than an existing technology which they already own. Generally this can be prevented by an 'obligation to use' clause in the contract between a non-employee scientist and the corporate sponsor, but this would not apply to in house research.

From the above discussion, it is clear that the ethical case for patenting, on the grounds that it is necessary to encourage the public dissemination of knowledge, needs some qualification. It is by no means a foregone conclusions that this will always be the result. Nonetheless, for all the shortcomings of the system, this probably remains the strongest ethical reason in favour of patenting biotechnological inventions generally, albeit not necessarily any and every invention.

Consequential Arguments against Patenting of Genetically Modified Organisms

The two arguments above represent the major ethical case for the patenting applied to genetically modified organisms. Various negative consequences have also been pointed out, and these are now discussed.

Patenting Encourages the Disadvantageous Features of Genetic Engineering

This expresses a broad intuition that the patenting of living organisms encourages the human exploitative treatment of the environment in general and animals in particular, and is not the right course of action. This is largely covered by the argument of principle discussed above, but there are some specific results which have been suggested. Some environmentalists have gone as far as to argue that our respect for life and its diversity is subverted so much by setting up a system of patenting of living things that ultimately we risk endangering our future survival. Many would see this as an exaggerated fear, however.

Some fear it would encourage cruelty to animals, because the profit motive encourages more work to be done on producing transgenic animals.[213] One argument against the oncomouse patent is that it is immoral to make money out of animal suffering.[214] Against these, it would need to be shown that patenting in itself makes these specific concerns worse. The profit motive is there regardless of patenting, and a patent is not a right to sell. Moreover, anyone engaged in production of genetically engineered animals will still be subject to national laws on carrying out genetic manipulation.

Patenting Can Have Adverse Effects on Those Involved in Production

If there is a good ethical argument for the place of the inventor, thought must also be given for effects of the patent on those using the invention, and whether it will mean there are losers as well as winners. Patenting could potentially place additional burdens on farmers through payment of fees for use of the patented organism. Obviously, there will only be a market for such products if they give at least some farmers an economic advantage, but this could then raise a question of equity. There is a concern that genetic engineering may only be of benefit to the rich producers who can afford to invest in its products. If the price for the continued use of patented seeds, plants or animals puts it beyond the reach of smaller farmers, this could increase the inequality between rich and poor farmers, and, on a national scale, between rich and poor nations.

Patenting can also be a key element in the exploitation of genetic resources of developing countries. Many people have expressed the fear that it is allowing commercial companies to take genetic material from a developing country and either fail to recompense them for it, or even, in the limit, claim invention, and charge them for reusing it, which would be grossly unfair to the country of source of the gene. These questions are discussed more fully in Chapter 9.

It is important also that some form of farmers' privilege is provided, allowing them to resow seed or breed from stock in the following years. This is needed to avoid creating a new kind of tenant farmer who has to pay royalties on every offspring of genetically engineered species, where the ownership of seed or stock effectively no longer rests with farmers but with biotechnology companies. From a point of view of justice, the intellectual property rights of the inventor have to be balanced against the effects on the producers downstream.

Patenting Can Have an Adverse Impact on the Scientific Research Culture

Patenting seems to fit better with industry, where the goal is a marketable product, whose competitive context implies secrecy, than the academic and non-industrial sector, whose culture sits uncomfortably with the concept. It is often difficult to identify the commercial potential of basic research before publication, and harder to control the release of information in non-industrial research laboratories. The high costs of protecting and defending intellectual property rights may be beyond the means of the laboratory. The proper delineation of research effort between public and private sectors is complex and highly politicised, but it raises the question of whether the implementation of a patenting culture in the public sector is beneficial or detrimental to the practice of science.

In 1992, the UK Government took the view that the technological future well-being of the country, its international competitiveness and its ability to exploit its intellectual base for wealth creation are intimately dependent upon making full and increasing use of patent protection.[215] This has led to a much greater emphasis being placed on scientists and organisations who are funded by the public sector exploiting the commercial potential of their new ideas. One aim has been to counter what was seen as 'the, generally false, perception that the proprietary nature of intellectual property protection may restrict academic traditions of open exchange of research information and that research carried out using public funds should be freely available to benefit society as a whole.' There is thus a strong philosophical element in this emphasis, beyond the pragmatic aim of making academics more aware of the commercial potential of their work. The benefit of society as a whole is seen primarily in terms of the right to exploit commercially. This amounts to an intention to 'change the Scientific Culture'.

To many both inside and outside the research community there are elements of this which are disturbing. Government sponsorship of research is increasingly reduced to customer-contractor principles and an internal market. These are far removed from the ideals of research which most scientists were brought up on, and which motivated their going into science, rather than say commerce, in the first place. Scientists may well resent the secrecy which patenting demands for otherwise exciting new findings or publishable results. Again, this may have been one of the reasons why they did not go into industrial research. In a situation where research funding is much harder to obtain, if the criterion for obtaining further grants depends largely on the number of patent applications put in during the previous year, it will inevitably skew research goals away from the logic of the science for a short term return. It is one thing to be fully aware of the side steps of a programme that will lead to commercial products, but quite another to drive the research based on them.

In the new culture rewards and incentives to scientific staff now take greater account of success in exploiting intellectual property. If patenting became the prime criterion for gaining financial support and personal career advancement, what sort of culture of research and researchers would that create? Confidentiality, distrust of competitors, hushing up inconvenient data, cutting research corners are apt to be seen as appropriate values for commercial success, instead of the traditional marks of objectivity and self-criticism amongst scientists. Baser motives have always been present, but the implicit tying of research to the profit motive suggests an undue place of mammon instead of truth.

These policies are supposed to address the popular belief that British scientists are brilliant, even unrivalled, at inventing things, but second rate at exploiting them commercially by comparison with other countries. While there seems to be some truth in this, it can be argued that the crucial barrier lies much less in the attitudes of the scientific community, and more in the short-term orientation of the financial community. When applied to the returns on biotechnological research, this approach is myopic. In the long run, it is likely to result in a diminution of the national research base, not just for future innovation but future knowledge. A person's life does not consist in the abundance of his or her possessions.

POINTERS TO THE WAY AHEAD

Is Patenting an Inappropriate Mechanism to Use?

The various problems posed by applying patenting to genetically modified organisms, and animate matter in general, raise the question of whether patenting is the most appropriate mechanism to protect biotechnological intellectual property. It is simply the system currently in use, but sufficient problems have arisen in adapting it to living matter to suggest that it may be the wrong approach. To take a system largely shaped by industrial manufacture, and force living organisms to fit into it, seems to be approaching the question backwards.[216] It would seem better to have gone back to first principles and ask what system of intellectual property rights would best serve the case of biotechnology, given that living organisms are involved, which have a different moral and ethical status to objects, machines and chemical processes. Would some adaptation of breeders' rights be better, for example, or something entirely different?

The report of the UK National Consensus Conference on Plant Biotechnology in 1994 made an important observation about patenting. This was that current patenting procedures appeared to them to be both a risky, technically inappropriate and an inadequate method of dealing with the issues raised by plant biotechnology. The lay panel at this conference recommended that this issue should be examined in an appropriate legislative forum, with a view to providing a more appropriate framework either completely new, or adapted from another existing system.[217] Despite the passing of the EU Directive, where an opportunity was missed, this remains a question which calls for examination.

The Need to Relate Patenting to Public Ethical Assessment

A further problem arises over how patenting (or any other system which might be proposed) relates to the overall way in which genetic inventions are made, presented to public scrutiny, and go on to be marketed. A patent is not a licence granting permission to do something but to restrain others. Granting a patent, however, implies that someone, somewhere, has already made an official evaluation of the ethical and social acceptability of the invention. If it were unethical, strictly speaking it should not have even reached the stage of patenting. In practice, the invention may well have had little or no ethical assessment of this kind. This brings to light two important questions which are currently ambiguous:

- At which point (and by whom) in the process of discovery and reporting of an invention should the definitive assessment be made of the ethical acceptability of the invention itself? Even if the original experimental work had to pass through an institutional ethics committee, that is not the same thing as assessing the ethics of the final marketable invention.
- How does the refusal of a patent on ethical grounds relate to the legislation which allows the invention to be produced at all?

At present an anomaly appears to exist whereby a patent can be refused on ethical grounds, as with a mouse used for research into baldness, but there need not be an equivalent ethical ban on producing the invention. Ironically, because patenting is a negative property right, if a patent is turned down on ethical grounds, it does not mean that no one is allowed to make the invention, but rather that *anyone* is free to make it! If there are to be ethical grounds on which a patent can be refused, these clearly need to be linked to an automatic ban on marketing the invention.

This would put an undue onus on the patenting system to act as ethical assessor on the public's behalf. It would therefore make more practical sense if there was a mandatory, parallel procedure of ethical assessment of inventions – perhaps a statutory national ethical commission, or an ethical assessment office analogous to the Patent Office. Any invention for which a patent was sought would at the same time have to go through this parallel procedure, which would include the power to ban it on ethical grounds. Like the environmental release assessment, to avoid a complete log jam of applications, there would probably need to be some fast track system for inventions which are very similar to ones already passed, so that only the more ground breaking innovations would need a full, in depth assessment. Like patenting, it would need to have right of public comment. The invention would be made public, assessed and given interim intellectual property protection while opportunity is given for the public to lodge objections (or voice support) and to appeal.

The European Parliament originally asked for a committee charged with making ethical assessments of at least the more sensitive inventions for which a patent application had been made. The European Commission refused, on the grounds that this would 'disturb the patent procedure'.[218] It reduced the demand to a general reference to the existence of the Commission's Ethical Group on Science and Technology, just at the time when the remit of this group had been extended from biotechnology to all technology. Moreover, their function is to make only general assessments of broad issues which, together with their slender resources, precludes any viable role in providing the ethical assessment of particular inventions. In the light of the increasing European public concern over genetic engineering, this judgement and the view that ethics must not disturb commerce could be regarded as a remarkably retrograde step, where the commercial factors are seen as the supreme criteria in practice.

What this highlights, once again, is the absence of a consistent public way by which inventions involving genetic engineering are assessed ethically in public. Part of the present problem over patenting genetic inventions or animate matter is that the publishing of the patent is often the first time members of the public have any way of knowing about the inventions, and, in particular, have effective access to an appeals procedure. This inevitably puts the ethical focus on patenting, which is the wrong place. It is clearly unsatisfactory that, in the context of patenting, an examiner in the European Patent Office is required to make a determination of the ethics of each biotechnological invention without recourse to wider public consultation, and yet for there to be no effective provisions in UK or European legislation to decide the ethical acceptability of a particular invention itself, nor for public consultation on the matter. Some solution is clearly needed. The above suggestions are not meant to be definitive, but rather to illustrate what would need to be achieved by whatever system was eventually developed. If something is not done to grant such a public debate, the controversy over patenting living organisms will continue without any hope of resolution, and, in the long run, the bioindustry risks losing part of its public for what may be the wrong reasons.

CONCLUSIONS ON PATENTING AND GENETIC ENGINEERING

A fairly strong ethical case can be made for some form of intellectual property protection relating to the area of genetic engineering, on the grounds of basic justice, of protecting investment and of encouraging the open dissemination of commercially sensitive knowledge in a competitive climate. It is less clear that it is an absolute prerequisite to safeguard the continuation of genetic research; it is more to do with national competitiveness. Each of these ethical goods, however, also carries with it some significant problems in the way the system operates in practice, which at least dull the edge of these positive arguments. For example, both in terms of justice and protection of investment, the system tends to favour larger companies and those richer farmers who can afford to buy the more sophisticated products. Smaller biotechnology organisations and small farmers may well lose out, relatively speaking. Similarly the open publication argument is compromised by the many problems which exist with regard to the information flow which patenting is designed to enable, and by the abuses of the system which can restrict information. The effect of a patenting culture on academic and non-industrial research has a number of serious disadvantages, not least on the scientists themselves. Some of the more sweeping claims made for the benefits of patenting are therefore an exaggeration, but there is clearly a good overall case.

Notwithstanding the passing of the EU Directive, doubts remain about whether it is the right approach to adapt the historical system of patenting, making a rather awkward accommodation for living organisms and genetic modifications. What is more contentious is what should be patented, or otherwise protected, and what should not. Much depends on whether it would provide adequate protection simply to grant patents on either the process of modification, or the application of a modified organism to solving a particular problem, instead of the organism itself. In view of the ethical sensibilities of many people, mere convenience to the inventor would not constitute grounds enough for insisting on the latter.

Many people hold strong principled objections to the patenting of living organisms in themselves, whether genetically modified or not. The main grounds are that invention cannot be claimed for a living organism which God has created via natural selection, and that patenting blurs or even removes the normal distinction between animate and inanimate matter. An animal does not constitute an invention or a process, in the normal sense of either word. There is a case to be answered whether patenting is contributing to a degrading and overall attitude towards God's creation, because the way in which we classify things also affects the way we treat them. Others disagree and say that all these matters are peripheral to a patent. What is important is that they are addressed in the implementation of the invention. This is complicated, however, by the question of association. As far as the public is concerned, granting a patent may be seen as representing an acknowledgement that the development is ethically acceptable as well, whether this is true technically or not.

There are serious deficiencies in the way that patenting relates to the ethical assessment of genetic and biotechnological inventions. Patenting is only a negative property right and is not designed for major ethical assessment. Yet it often ends up as the place where such discussion is focused because of the absence of any other suitable forum to which the public has ready access. There should be some mandatory parallel system such that any biological invention put forward for patenting must also undergo an ethical assessment for its acceptability as an invention in its own right. The failure of the EU Directive on biotechnology patenting to provide adequate provision in this respect represents a serious setback for the wider public discussion of genetic engineering.

9 GENETIC ENGINEERING AND DEVELOPING COUNTRIES

FOOD AND POPULATION

One of the biggest single problems facing the world today is the growing imbalance between an expanding world population and a static, or only slowly expanding, world food supply. The problem is made more complex by the fact that the developed world produces about two thirds of the world's total food supplies and has a fairly static population which is well fed, perhaps even overfed. In certain areas such as Europe and the US, agricultural land is taken out of food production as set aside, or land bank, in order to control the over-production of food. In contrast the developing world has a rapidly growing population, due to improved survival rates, especially in infancy, yet its food supplies are not growing at a proportionate rate. Much of the potential arable land is currently being taken out of production, or is deteriorating due to desertification and deforestation, or is being turned over to cash crops for sale to the industrialised world.

The current world population is about six billion and rising. Although there are some signs that the increase will not be as dramatic as had been predicted, it is likely to rise inexorably to 8.5 billion in the next generation. It is estimated that the agriculture of the world could supply enough food for 10 billion, and perhaps as many as 15 billion with better distribution and sensible prices. Even this level is based on the assumption that the political and economic factors which hinder world trade in food could be resolved, an assumption which many think is too optimistic. However, there is no known way for the world's agricultural land to support more than a maximum of 15 billion people. Issues of food supply are complex and affected by a variety of other factors which may be unrelated to production capabilities, such as wars and the effects of economic systems, but these are beyond the remit of this book.

It is argued that one of the most important pre-requisites to achieve reasonable birth control is to enable developing countries to afford a better standard of life, since only when people's lives and livelihoods are assured do they become willing to limit their family size. To many people in developing countries today, a large family is an insurance for old age, something to be sought rather than avoided. Many economists believe that it is essential for developing countries to develop their own economies as fast as possible, including some reasonable application of biotechnology. However, this is a contentious point, with other economists pointing out the wider costs which this often carries.

TWO APPROACHES

The problem of feeding the world typically provokes two (often opposing) analyses as to how the problem should be posed, and two approaches for addressing it.

The first view is that there is an urgent necessity to deploy the resources of biotechnology to the problem, for without this technology we will not be able to feed the future population. This approach concentrates on finding a technological solution to the problem of increasing food supplies. In essence, it is argued that by understanding more about the biology of plants and animals, the techniques for growing food can be improved so that more food can be produced. Genetic engineering is another tool for helping to do this, and with the right application could be a very powerful one. In support of the usefulness of this approach, various success stories of the green revolution are cited. For example, in parts of Malaysia, Latin America and a few countries in Africa there has been a move towards systems of agricultural production typified by the use of modern hybrid varieties of maize and sorghum, the use of imported exotic breeds, the development of stable crossbreeds, or the use of machine powered methods of agricultural production (for example the tractor rather than the draught animal).

The second approach argues that the assertion that genetic engineering will help solve the problem of world food supplies is made with no reference to the social context in which the technology would be applied. This approach seeks to solve the problem of food supplies by looking at the sociological context in which agriculture is practised, giving more priority to indigenous knowledge and to issues such as changes in land ownership and power structures. It is argued that the transfer of technological agricultural improvements from the industrialised world is unsuited to the real needs of the developing world. Proponents of this view point to some of the problems arising from the green revolution. For example, the high input requirements of improved varieties have led to dependence on imported seeds and fertilisers and in some cases have resulted in problems of pollution. They have also encouraged the development of large plantations and the inequitable distribution of the green revolution's benefits.

DIFFERENCES AMONG DEVELOPING COUNTRIES

While the term 'developing countries' is used in this document, it is important to recognise the huge variety among the countries which are grouped under this umbrella. The impact of genetic engineering is likely to depend very much on the situation of an individual country. James and Persley[219] considered there to be approximately 12 advanced countries (eg China, India, Brazil, Mexico) which have national programmes with the capacity to work in biotechnology, and approximately 100 countries (eg Ethiopia, Bolivia, Burma) which would not be able to incorporate biotechnology without a considerable amount of help. For the poorest countries, there may be a limit to their ability to embrace any modern agricultural technology, let alone genetic engineering.

In the more advanced developing countries, on the other hand, the improvements in productivity which genetic engineering can offer are likely to be accepted and implemented readily. The advantages are generally perceived to outweigh the risks by far. Ethical constraints on their use may not be regarded as being of major consequence in many of these countries. Such factors would seem to be far less important, in ethical terms, than the imperative of economic improvement, or perhaps the notion of advancement represented by the technological practices of industrialised countries. For instance, the animal welfare lobbies in many such countries have little influence and are regarded by many as strange imports of weird ideas from foreign countries. Some politicians have even gone as far as arguing that considerations of animal welfare and environment are arguments adduced by malevolent foreign countries to discourage development and thus perpetuate some form of neocolonialism. Such perceptions will also vary from country to country, and they may well change, but at the present time ethical constraints are likely to be far less than those prevailing in the industrialised world.

CASH CROPS

So far this discussion has centred on the need of developing countries to become self-sufficient through increased food production, but a great deal of the agricultural land in many such countries is planted for cash crops and not for food. Thus there is more land planted to palm trees in Malaysia than to all the food crops combined. Similarly, until recently in the West Indies the acreage of sugar, cocoa, coffee and coconuts far exceeded the acreage of staple crops and vegetables.

The growing of cash crops for export is one route in the drive for developing countries to increase their economic well-being which, as discussed earlier, is seen as a pre-requisite to improving their standard of living to an extent where the practice of birth control is a real and not a theoretical option. Cash crops are not the only means of improving the national economy, but they have been used increasingly in order to maintain debt repayments to international banking systems.

Where cash crops are a vital component of a country's gross national product there are two real dangers. Firstly, concentrating on them may reduce the country's own ability to become self-sufficient in food. A good historic example is the growing of cloves in Zanzibar, where the sultans decreed that local food production had to be curtailed to make land available for the growing of cloves for export, on which taxes could be imposed. Secondly, an over-emphasis on cash crops could attract foreign entrepreneurial companies to monopolise the best agricultural land for large scale production, as in the case of the sugar estates formerly occupying large areas of land in the West Indies and parts of South America, or the tea estates in East Africa and Sri Lanka. Thus land, labour and capital would tend to be attracted away from other areas of agricultural production.

This consideration is relevant to genetic engineering in two respects. Firstly, it is argued that these cash crops (rather than food crops) are the most likely targets for genetic engineering, since they are more able to generate an income for the innovators to recoup their investment in research and development. Overall, this is not likely to benefit food production for developing countries.

The second issue is that genetic engineering is in some cases being applied to plants and other organisms in industrialised countries to produce cheaper substitutes for imported crops from developing countries. An example is the modification of oilseed rape to produce oils similar to those from the oil palm (Case Study 3). The sudden collapse of a country's main export market, due to substitution by a western genetically engineered product, could have a catastrophic impact on that country. It might be argued, at least in theory, that land so freed from having to produce cash crops could then be used to produce more food, but if the overriding need in the country was for hard foreign currency, this would not necessarily solve the problem, and might make things worse.

INTELLECTUAL PROPERTY RIGHTS

One of the most contentious issues raised in connection with developing countries is that of intellectual property rights, particularly patenting, and the associated question of maintaining biodiversity. Intellectual property rights are explored more generally in Chapter 8, but their application to developing countries merits special discussion here.

Historically, few countries have relied solely on their indigenous crops for feeding themselves; for example, a United Nations report[220] suggests that many major Canadian wheat varieties contain genes from as many as 14 developing countries. This is not restricted to industrialised countries; developing countries are also dependent on each other. For example, bananas are the most important cash crops in South and Central America. The highest per capita consumption of these crops as staple foods is in East Africa, but the origin of bananas and plantains is in South East Asia.[221] There is thus a complex web of interrelationships between countries in terms of their genetic resources.

There are three main issues raised by intellectual property rights in relation to developing countries.

The Exploitation of Developing Country Genetic Resources

There have been several examples of western scientists or commercial organisations obtaining genetic material from a developing country, incorporating this in some new use and patenting the resultant product, so that the country of origin is required to pay to purchase back the new product based on their indigenous plant. This is manifestly unfair. An often quoted example, which does not involve genetic engineering, is that of the Neem tree. Neem seed extracts have been used in India for generations as a source of insecticide in agriculture, for treating eroded and infertile soils and for certain human health problems. A US company has patented a process by which the oils can be extracted and fractionated to give a more consistent product with more predictable properties. The company has had to carry out research in order to develop this product and therefore looks to make a profit from it. However, the concern is that the indigenous people who had conserved knowledge of these properties have not benefited at all from this development.

The problem is exacerbated by the fact that most seed banks, where samples of numerous developing country plants are kept, are largely in the industrialised countries. The danger is that, in practice, they are controlled by industrialised nations. Thus, developing countries fear that they have no control over the way this germplasm can be exploited. On the other hand, commercial companies point out the large research investment which is often needed to get from the plant to the product and that, most importantly, many developments are not successful. In the pharmaceutical industry, a success rate of one in ten is regarded as high. If the developments that did work could be copied by anyone, the companies claim that they could no longer generate sufficient income to carry on their long-term research and development work.

Under the Uruguay round of the General Agreement on Tariffs and Trade (GATT) negotiations, signatory states are obliged to adopt a patent system for micro-organisms and to adopt some form of intellectual property rights for plants, either as patents or some other form, for example UPOV Plant Breeder's Rights (see Chapter 8). This action has focused attention on to the issue of intellectual property rights and has resulted in widespread protests in India, where farmers feared the large rises in the prices of seeds and an attack on their traditional rights of saving seeds from their harvest.[222] Although there is the potential for this to occur, it remains to be seen whether this will occur on a widespread basis. Others argue that, with the adoption of the GATT requirements, the only intellectual property in the world that would not be protected would be that of indigenous communities.[223] There may be particular circumstances where it would not be desirable for farmers to keep back their own seed for planting. An example would be that if a cross-breed of two varieties was being used, segregation would take place in the next generation and few of the progeny would be of the same type as the parent generation.

Different Approaches to the Notion of Ownership

Among the countries of the world, there are several different approaches to the notion of ownership. The concept of intellectual property rights requires an acknowledgement of individual ownership of most products and processes. In many countries, essentials such as food crops are viewed as owned communally. Current intellectual property protection systems are ineffective in supporting community level innovation.[224]

Diminishing of the Gene Pool and Knowledge of Indigenous Crops

Concern is also expressed that the production of genetically engineered plants with improved production characteristics will mean that these crops will displace large areas of indigenous crops. The result will be a diminution of the crop gene pool. It is also argued that this may result in the loss of indigenous breeding knowledge, particularly with respect to plants adapted to particular ecological niches.

Attempts have been made to overcome some of these problems. A notable example is the bilateral agreement in which the pharmaceutical company Merck paid Costa Rica for the rights to biodiversity exploration, and also committed itself to training workers and to paying a royalty if any commercially attractive products were developed from these resources.[225] Proponents argue that this is an example of a constructive way forward. Others have argued, however, that the sum paid by Merck was far too small.[226] Additionally, it is stated that if such bilateral agreements proliferate, there will be countries who will be left out altogether. While such a scheme may not be perfect, however, it seems better than offering no recompense to the country or community which is the source of the germplasm. With time, it will be possible to review how well such schemes have functioned.

In Chapter 8, the case is made for a more appropriate framework than patenting to cover intellectual property rights involving living organisms. Without the benefit of some such rights, there would be a reduced incentive for breeders to produce better varieties, but it seems clear that there will be many cases where the present system of patenting will be disadvantageous to developing countries.

RISK

The issue of risk and genetic engineering is addressed primarily in Chapter 7, but three features are worthy of mention here. Firstly, products tested in western countries may be used in developing countries, without equivalent levels of testing for different local conditions. Many problems could potentially result.

Secondly, tight regulations prevailing in richer countries of the world may encourage companies to export their product development to a developing country, where regulations may not (yet) be so strict and where enforcement may not be so well applied at present. A notable example is trials of a recombinant rabies vaccine in Argentina, apparently without the government's knowledge.[227]

Lastly, the developing country context can radically change the usual circumstances in which risks are evaluated in richer countries. The relative risks from different courses of action are especially altered in emergency situations. Where the need is for the basic necessities of life – food and water – it would appear to many to be academic how cleanly the food and water are produced. As has been seen in Sudan, Somalia and Rwanda, even grossly contaminated supplies of food and water will still be utilised in dire circumstances when demand far exceeds supply. Normal constraints about quality, safety and efficacy then take second place to mere survival.

In less extreme circumstances, there may also be pressure to set western risks aside, because at least a worse risk is being addressed. The use of rBST to increase milk production (see Case Study 7) has certain drawbacks and has not yet been licensed in Europe. It has been approved for use in, for example, Zimbabwe, Mexico, Brazil and Pakistan, where it might be argued that the need for increased production could outweigh the problems. It remains to be seen how extensively it will be used in these countries in reality. China led the world in releasing GMOs over a much wider area than anywhere else, apparently because the benefits outweighed any fears about risk. A Chinese proverb states 'If a man has food, he has many problems. If a man has no food he has only one', but even so this is not *carte blanche* for inappropriate solutions.

WHO SETS THE RESEARCH PRIORITIES?

Buttel[228] points out that, during the green revolution, national research into agriculture was relatively well funded. This is no longer the case, and agricultural research has become increasingly privatised and profit motivated through such political initiatives as 'prior options'. This exercise in the UK has led to the examination of all the research institutes with a view to identifying those which could be privately funded. The fear is that privatisation will result in a concentration of research into profitable areas and/or high technology projects like genetic engineering, which may well not be of significant benefit to developing countries. It is argued by some that such research would tend to benefit richer farmers to the detriment of the poorer farmers.

Furthermore, analyses of technology transfer programmes show that many well meant ideas are doomed to failure. Often this is because of an unwillingness or inability to understand fully the context in which the new technique is to function. This has been true of appropriate technology projects as well as more conventional ones.[229] The traditional 'top down' approach is seen as less successful than a participatory model in which farmers are involved throughout the process. The belief that genetic engineering has the potential for improving food production does not necessarily mean that this desirable outcome will be achieved.

CONCLUSIONS

Where does this leave the two approaches which were spelled out at the beginning? It is not so much a case of choosing one, but of picking the best points of both. Economic growth in many developing countries is much to be desired. But there is no point in achieving this by copying some of the mistakes of the richer countries, nor in applying inappropriate solutions based on sophisticated 'high tech' methods, where something more participative would work far better. A start has been made on genetic engineering applications aimed at the richer countries of the world. There may well be different genetic engineering developments which would benefit the developing world immensely, but these have not emerged so far. If society is serious about this aim, now is perhaps the time to begin to ask what would really help most, and endeavour to secure a positive contribution to solving the immense food versus population problems of the developing countries.

In balancing the two approaches, the industrialised countries have a responsibility to ensure that the biological resources of developing countries are not unfairly exploited. Two key features of this are securing stable funding for bodies set up to do research appropriate to developing countries, such as the international agricultural research centres, and ensuring that developing countries have adequate representation on any international forums on biotechnology, including addressing issues such as patenting. These recommendations echo those made by the Report on the UK National Consensus Conference on Plant Biotechnology in 1994.[230]

10 THE SOCIAL CONTEXT OF GENETIC ENGINEERING

INTRODUCTION

The German sociologist Beck-Gernsheim poses a highly pertinent question:

> *Is there a path that will lead us out of the dictatorship of the experts and counterexperts, out of the expertocracy where the doctors, geneticists, biologists, sociologists, psychologists, jurists, and specialists in ethics lay out their pros and cons before us in many loops of argumentation, while all of the others, the people in the outside world, trust that the experts will know where they're leading us or simply follow the course that seems plausible at a given time or happens to be in the headlines? In other words how can we keep democracy alive in this area – in decisions that are so fundamental and that determine the course of the future?*[231]

Throughout this study, the social dimension of the issues raised by genetic engineering has been a recurring theme. It has proved very important to appreciate the different social constructions of the questions, and the societal situations of the different players involved, in order to gain a proper understanding of the moral debates themselves. It has also emerged that there are serious shortcomings in the level of societal involvement in the decision-making being done on the society's behalf. While genetic engineering has many novel features by virtue of the innovating character of the science, it represents a striking contemporary example of the interplay between science, technology and the structure of industrial societies. This chapter explores those interactions and finds that they throw valuable light on the issues under discussion.

Firstly, it stands back to identify certain aspects of the nature and practice of science which are key in the subsequent discussion. The idealised notion of pure science is seldom a fair representation. It then illustrates how the wider social context in which science is practised provides certain values. These values in turn affect how science is translated into technological change, and in particular how the paradigm of genetics has been applied. The questions of who owns, who benefits and who controls emerge as crucial issues. Three important questions are examined which are key to weighing up the claims and counterclaims of the value judgements about genetic engineering. The first is the existence of different forms of rationality, and the role these play in shaping the resulting conflicts over genetic manipulation and the various interest groups involved. Emerging from this comes the vital issue of control and accountability. Finally we consider the way risk is understood by different sectors and the role it plays in society.

SCIENCE AS ORGANISED SCEPTICISM

There are three particular characteristics of science which are relevant to this discussion – the questioning of myth, provisionality and effectiveness. Science, as a practice, can be seen as a special form of organised scepticism. In order to obtain knowledge about the world, problems which once were only considered in terms of myths are explored in ways which are testable. Thus, notions based on received truth alone are challenged. Traditional stories about the origins of things and the structure of the universe are not simply passed on as dogma, but subjected to questioning. Theories are constructed and put to the test in terms of their logical coherence and empirical evidence. This critical attitude was seen as the basis of scientific progress, the birth of a new tradition. In doing so, however, science in its turn has not escaped becoming the subject of myth, as will be seen.

A corollary of this critical approach is that, according to some philosophers of science like Popper, scientific knowledge itself is always provisional, precisely because it is subjected to systematic doubt. The working paradox is that while science enables us to obtain knowledge of the real world in place of myth, its own inherently provisional nature makes its findings subject to uncertainty. Scientific activity can be seen as a process of conjecture and refutation, in which the process of falsification is crucial. According to Popper, if a theory cannot be constructed that is capable of refutation, however interesting it might be on other grounds, it is not scientific. It is through such falsification that science progresses.

This is at the heart of what he termed critical rationalism. It is to science what the Socratic method is to philosophy.

Yet despite the provisionality, science is not just one form of knowledge which has no more significance than any other. Even if science cannot make absolute claims, neither is its knowledge merely relative. However provisional its theories may be, the power of scientific knowledge rests in the fact that to a very large degree, and in so many applications, it clearly works. Planes fly; vaccines are effective; bridges get built. Indeed, when an application does not work, science is not abandoned. Rather, the reasons for failure are enquired into, and mistakes rectified. There is a process of development involved, which is characteristic not only in the growth of knowledge but also in the way science is applied to the world. In the process, however, it also produces the logic of the 'technical fix' for the problems which science generates, as opposed to stepping back and thinking if the application was wrong.

THE SOCIAL CONTEXT OF SCIENCE

Science is not a detached activity, running independently according to its own rules. There is a social context in which scientific activity takes place, which includes many wider cultural considerations about ways of life, and the things that are held to enhance or to diminish them. The reasons science is applied in one way but not another, and the selection process, which leads to some applications being given priority over others, are both indicative of questions of value, ethics and politics, which arise from a particular social context. This is most obviously the case for technology, but it is also true of science itself.

The Historical Religious Context

The emergence of modern scientific activity is sometimes identified as a conflict between science and religion, especially given the challenge to myth and received truth. The nineteenth century controversy over Darwinian evolution symbolised in the debate between Bishop Wilberforce and Thomas Huxley is a much cited but misleading example. The perception of conflict is at odds with the context in which modern science arose. Among many scientists in the seventeenth century there was a motivation for their work grounded in an attempt to explore nature in a rational, systematic and empirical way, in order to glorify God and His works – 'to think God's thoughts after Him', as Kepler boldly put it. The setting up of the Royal Society in the mid-seventeenth century was done with the active support of religious leaders, often Puritan divines, and scientists with strong Christian commitments, such as John Wilkins, Robert Boyle and Sir William Petty. Francis Bacon, whose work influenced the principles on which the Royal Society was constituted, referred to scientific activity as being concerned with the glory of the creator and the relief of man's estate, a combination of concern with discovery and the application to human welfare. In other words, this is technology.

In his seminal paper, *Puritanism, Pietism and Science*,[232] Robert Merton concluded that there was a clear association of Protestantism with scientific and technological interests because of the norms which both embodied. There was an affinity in their individual questioning of accepted authority, in the right or even the duty of freely examining the world around, in the value of an empirical approach, and the drive to study nature so that it might be controlled. The direction of early modern science derived significantly from these roots. In due course, the religious basis was displaced and gave way to a view of science seen primarily in utilitarian terms, without the framework of ethics which had derived from Christianity. What we now have to come to terms with is that, despite its skill in identifying causality and moving away from myths, having demystified the world, science has no values of itself. As Weber observed at the end of the First World War, science does not tell us how to live.[233]

The corollary of this has been not that science and technology now run autonomously without values, but, as observed in Chapter 3, the implicit values by which they run have become invisible, often even to their practitioners. One of the purposes of this book is to re-establish the need for explicit values in biotechnology, and where appropriate to present alternatives to the prevailing largely utilitarian evaluation – pointing in particular, but not exclusively, to the worth to be gained from Christian ethical insights.

The Scientific Community as a Context

This highlights the importance of the scientific community itself. Especially since the work of Thomas Kuhn, scientists have been seen not just as individuals but also as communities.[234] Their knowledge in a particular area comes to be developed so that, as a community, they practice 'normal' science, according to the current understanding. Their work and their criteria for testing operate within that paradigm. A scientist may begin to think that new evidence coming to light requires breaking out of the paradigm. But it not always easy to do so, given the professional controls that govern the community, including the ability to publish in the scientific literature. Eventually, however, the lack of fit and the presence of anomalies in the present framework can lead to a paradigm shift that can sometimes be so significant as to be described as a scientific revolution. While not necessarily agreeing entirely with Kuhn, it can be recognised that it is the maintenance of the critical spirit that can eventually overthrow the conventional scientific wisdom of received orthodoxy.

Genetics as a New Context

Crick and Watson's discovery of the structure of DNA heralded a new paradigm which has had just such a revolutionary effect. It has been a revolution not only in the understanding to which the resulting 'new genetics' has given rise, but also in the far reaching consequences of the applications which have become possible, and particularly genetic engineering. Genetics has provided a paradigm for an entire new vision of human potential, and the potential of nature. Within this vision, the very term genetic engineering is deeply significant. It borrowed the idea of engineering from human mastery of inanimate materials and applied it to life forms, through the harnessing of the new skills and understanding given by the genetic paradigm. Now on the basis of pure science it has become possible to work with materials, whether animate or inanimate, disaggregating things that were formerly connected, reassembling them with other materials so that something new is constituted with hitherto unavailable possibilities. There is a breaking down, manipulation and the creation of new assemblages. Linguistically, the ruling concept is the mere application of skill, a utilitarian perspective, and this has blurred the value distinction between animate and inanimate and what we should or should not be permitted to do to either of them.

The Technological Context

Genetics is a specialist branch of science which is permeated by technological concerns. The earlier reference to Bacon is a reminder of the powerful interconnections that science has had with the processes of industrialisation since the seventeenth century. The concept of pure science has stressed as its core value the idea of disinterestedness – knowledge for its own sake – but in reality, concerns with applied science have never been very far away. It is here that extra-scientific questions come especially into focus. Wide ranging discussions are engendered about the relevance of science to society, in terms of new products and processes, and its significance for efficiency, productivity and human welfare, and, as has been observed, typically placed in a utilitarian framework of some kind.

Like nuclear power, genetic engineering has become a highly significant example of applied science in the second part of the twentieth century. Applications such as these generate hopes and fears, and are subject to evaluations with ethical, political and economic connotations. They become areas of contention, far from abstract notions of science as pure knowledge detached from the world and human societies. Indeed they have become embodiments of two contradictory modern myths about science and scientists. There is the myth of the all-powerful, all knowing scientist, who can guide us into a better future by bringing manifold new benefits and solving the problems that beset us like disease and hunger. There is also the Frankenstein myth of science which brings such changes into being that it puts us in thrall of powerful forces outside our control. It is ironic that science, which sought to supersede mythological accounts of the world, has not been able to prevent the emergence of powerful social myths of its own!

Science, Funding and the Commercial Context

Scientific research does not develop in a social vacuum. The research and development functions of universities, research institutes and industry-based laboratories represent different types of institutionally organised arrangements. Each is embedded in an organisation with a particular nature, purpose and aims, relating variously to agricultural, medical, industrial, educational, economic, political or military spheres. There may be many interests involved in why a particular field even of so-called academic research is pursued and funded, let alone more applied science. These interests might include those of pharmaceutical companies, seed manufacturers, cancer charities, government departments and the EU. There is also the scientist's own ambitions and pragmatic sense of what is likely to attract funding. Each organisation which has an interest in the outcomes of scientific research and potential applications will also have its own agenda and aims, and a set of values which these express. How this works out in practice, what is negotiable, what priorities are established with what funding, will give a strong clue as to the nature of the power relations that surround the activity.

It is generally recognised that research funding even in academic contexts is increasingly from the private sector, not least in genetic engineering. This is a shift brought about by changes in the level of funding given to universities and research institutes from the public purse, the product of a more market oriented approach within the education sphere itself. With this comes a shift in the nature of research and the values under which it is being done, and in particular a move to greater use of commercial values rather than the traditional scientific values of openness and discussion.

It has been argued that one of the core values of science is its universalism. Its truth claims must be subject to impersonal criteria and accessible to scrutiny in the public domain. This is far from always the case in practice, however. One aspect is connected with the issue of secrecy. What is not disclosed cannot be scrutinised. It has routinely been the case in areas like military research that the circulation of scientific research has been restricted or suppressed in the interests of secrecy, which themselves are grounded in the interests of the state. This is more blurred where the interests are commercial or political. It seems likely that the stronger these interests are in a piece of research, the greater the danger of less than open reporting and dissemination, either within the scientific community or outside.

This can raise serious questions of public accountability, if the research is dealing with far-reaching novel applications of results of genetic engineering. The term 'commercial: in confidence' applied to official documents is meant to be a reasonable safeguard. In practice it is apt to be used as an umbrella for whatever a company thinks is sensitive, and is wide open to being abused. Even for public information such as the notification of the release of genetically modified organisms, the information that is made available may be quite nominal, because the company concerned could claim it was commercially sensitive.

Patenting is put forward as a response to this dilemma, in that it is intended to provide for the open disclosure of information while protecting commercial investment. Despite the forcefulness with which this claim is often made, however, its effectiveness in achieving it leaves much to be desired, especially for the academic sector, for reasons explained in Chapter 8. Here it is simply noted that the ground rules for patenting vary across different societies and this itself is a reminder that these are socially constructed and contested, and not some sort of absolute.

Who Owns – Who Benefits?

The current trend which thinks of science increasingly in terms of intellectual property rights prompts one to think not in terms of the universality of science but rather of who owns it and why. With so much research being commercially driven, with corporate capital involved in financing applied science, the question 'who is benefiting?' also needs to be asked. Within a capitalist framework it is assumed that shareholders are expected to benefit economically as a reward for the risk they have taken. But if developments are profit driven, rather than consciously related to social need, then unless the market is treated as the arbiter of decisions to be made about human welfare (the celebrated 'invisible hand' of Adam Smith) questions of social justice and welfare become pushed to the margins.

Genetically engineered tomatoes and tomato paste represent a case in point (Case Study 6). In neither case can it be argued with much conviction that these products have been driven by a social need. In the UK context, even the production advantage is not convincing for a tomato that is easier to transport for very long distances. While for many people Zeneca's paste may not pose insuperable ethical problems, one cannot help asking whether, in the context of real social needs of food in developing countries, something much better could not have been done with the effort and investment. There is an opportunity cost in such research and development to which a focus on first world market economic criteria appears to be blind.

A further example comes in the case of genetically modified soya beans which is discussed more fully in Chapter 6. Despite widespread expressions of concern, these have been allowed into the EU without either being labelled or segregated from normal, unmodified soya. It has been suggested that considerable pressure was brought to bear by the commercial organisation concerned and the US government, to which the EU gave way, for fears of a trade war. There were company protestations that it was impracticable to segregate, or to label a product that would go almost everywhere through the food chain. This now seems to have rebounded somewhat, as some countries have insisted on such labelling, or else defied the EU altogether and banned the imports. It is again a case where the primary advantage is to the producing company, who sell the biocide to which the product has been geared. Initial evidence suggests that there are environmental advantages in that less chemicals are being used in the manufacture, but there is little other tangible advantage to the consumer. Moreover, any consumers who object on principle to genetically modified food or are not convinced of the arguments about environment or safety will now become losers. Even if most people do not so object, the society has a moral imperative to protect those who do, by labelling, and also to see they are not disadvantaged financially by having now to go to specialist food shops. This seems to be a case where the logic of the commercial sector appears simply to have overridden societal concerns.

The concept of the market is itself problematical. Those who point to its virtues from the standpoint of economic liberalism readily refer to the free market and the virtues of deregulation. Yet to all intents and purposes the concept of the free market is an ideological one, since by no stretch of the imagination does it describe the real world in which economic activity takes place. The world of corporate capitalism is one of very imperfect competition. The arena in which competition takes place is one where real flesh and blood struggles take place. Markets may on occasions be rigged and powerful groups can play their part in rigging it. In the newly privatised sector, many markets like electricity only work by virtue of being highly constrained by various government specified criteria. Regardless of whatever form it takes in practice, there is no possible basis for using the notion of the market as a sufficient basis upon which to make moral judgements about social welfare.

Nonetheless, in the wider field of food technology, the BSE crisis demonstrated the effectiveness by which on occasions people can bring their values to bear through what they choose not to buy. Consumer groups have in some cases mobilised their interests very effectively in ways which have affected the character of the market. This puts a question mark against models of technological or economic determinism. Hilary Rose has argued:

> *Molecular biology as a technoscience has enthusiastically embraced the market but it is learning, like other entrepreneurs, to be cautious about the resistance of the 'consumers' of its products whose opposition has been mobilised by the powerful movements and discourses of feminism and environmentalism.*[235]

Such considerations emphasise that scientists themselves are not innocent and detached players, given the context in which their work is done. If, given the kind of funding arrangements noted above, they are the servants of power, then it prompts further questions about the purposes of the powerful, since scientists are not masters in their own house. It is primarily the corporations that obtain patents for engineered crop plants, instituting intellectual ownership over what previously had no owner, namely the germ line of living creatures. One of the questions that this book has persistently asked is: for whom does this represent a loss and for whom a gain?

Technology, as a form of social practice, is never neutral. It is always connected with human purposes and interests. In a world of plural values it is not surprising that the question raised by applied science is not just how to do something, but for whom it is being done and why. For most current genetic engineering developments the context and purposes are set to a large degree by the market, and, as has been seen, this can pose problems in marginalising other values. This implies that some degree of intervention is necessary with the present system, if the whole of society is to have a realistic say in the inventions of genetic engineering. To claim, as some have, that we should not intervene with market forces is an ideological statement. Such forces are never untouched by human hand. The question is rather what kind of interventions can be undertaken, and how can they be evaluated. Is it possible to think in terms of the public interest and the public good, rather than mainly of private interests and the maximisation of profit? This then raises central questions about the role of politics and the democratic process, and how societal values relate to instrumental means and ends. It is here that the problem of rationality is confronted.

RATIONALITY, VALUES AND INTERESTS

In Chapter 3, reference was made to the perception amongst many proponents of genetic engineering that intrinsic ethical objections to genetic engineering were irrational, in contrast to their own rational assessment of consequences in a scientific and utilitarian perspective. A critique was made in terms of the hidden intrinsic values of the scientific perspective. Similarly in Chapter 7, it was noted that the precautionary approach to risk is regarded by some in industry and the scientific and political communities as irrational and non-scientific. The point has now come to examine more closely the notion of rationality, and the different meanings that are attached to the term, and the way it is both used and abused.[236]

Value and Instrumental Rationality

Max Weber drew an important distinction between what he called value rationality and instrumental rationality.[237] Value rationality refers to behaviour that is logically consistent with a particular value position which a person holds. Like inherent principles in ethics, in its purest form value rationality would not take consequences into account, as in the standpoint that rejects genetic engineering as intrinsically unacceptable. Nonetheless these absolute values are not always impervious to change in the light of new knowledge or experience. An example might be pacifism, which serves as a witness to a value commitment against violence, even though the value may not prevail in time of war. It is difficult to realise some values absolutely, like principles in the Sermon on the Mount, but nevertheless the commitment is made, and action is taken that is rationally consistent with that commitment. It is sometimes also described as the ethics of conviction.

In contrast, instrumental rationality takes account of the consequences of actions, as in utilitarian philosophy and various forms of cost-benefit analysis. It has to do with a calculative approach to problems and issues. The calculation relates to the appropriateness of given means to reach given ends. As a result it has a technical feel to it, in that the only evaluation which is of concern is the efficacy of the fit between the means and the ends. Its task is to clarify and attempt to establish the interrelationship between facts. The ethical consideration of whether we choose to endorse particular actions or policies has to do with our judgement on the desirability of the consequences. Weber also pointed out its drawbacks. Empirical analysis works only in weighing up the means, in so far as they relate to an unambiguous end.

In reality, however, the instrumental approach is not as straightforwardly rational as it might seem. As already discussed, the problems and issues of a technology are contextualised in various ways. There are different interests involved as well as different values. Thus genetic engineering has not appeared out of a vacuum. It has various stakeholders who each have implicit values in the ends that they seek. Moreover, for all its cognitive power, the role of scientific knowledge is not straightforward. Scientific rationality will not tell you what to do with its knowledge. It has to be interpreted and evaluated within extra-scientific frameworks with various ethical, political and economic dimensions. In an area as new and rapidly expanding as genetic engineering, there is an awareness that our knowledge of its consequences is currently quite limited. This awareness is itself a signal of how difficult it can be to practice the ethics of responsibility.

The key point is both of these are forms of rationality, but each uses reason on a different basis. It is not the case that by definition the instrumental approach is rational and therefore valid, and that the value approach is irrational. Nonetheless the myth is pervasive, almost ideological in some circles. Because reason has such powerful positive connotations, the notion of rationality has often become used as a tool of power in committee discussions or public debate. One group dismisses opposing views as irrational – usually referring to value rationality – in order to endorse its own position to marginalise the position of others.

Thus there are competing rationalities over genetic engineering issues, as one might indeed expect in a world which contains a plurality of values and a diversity of interest groups. This does not imply that every position is by definition rational. There may be logical mistakes on either side. In some cases it can be shown that people can be mistaken in their beliefs. Exposure to different viewpoints can lead us to review our own beliefs and assumptions. But given that there are these competing ways of being rational, the conflicts which inevitably arise have an important knock-on effect in leading to consequences which neither side intended. To this extent, the direction of genetic technology is socially as well as scientifically unpredictable, because of the outworkings of the social conflict it engenders.

Conflicting Rationalities within Science and in the Public

Various examples may serve to illustrate this. The first is from within the scientific community itself. In 1994, the US journal *Science* reported that over 30 leading scientists on the human genome project had made deals with venture capitalists. Tim Radford of the *Guardian* pointed out the alarm of European nations – who argue that research paid for by the taxpayer should be freely available to everyone – at the way some US scientists in the genome project had been 'furiously applying for patents on lengths of genetic code by the thousand, and the big companies have been taking an even greater interest in every aspect of their work.'[238] There can be conflicts within the scientific community, not only about the status of scientific findings, but also about ways in which they should be organised and deployed, and grounded in extra-scientific values and interests.

There is also an ambivalence within the wider public about science, quite apart from the few who hold a root and branch position opposed to science. The overall reactions to Dolly the cloned sheep (Case Study 11) have highlighted that, on the one hand, people are pleased at the prospect of new therapies and pharmaceuticals, but on the other, many are worried about the overall drift into innovations which challenge deeply held human values. Chapter 5 drew attention to another conflict between two cultures within society. There is enthusiastic endorsement of genetic research from the many emerging support groups for sufferers from various types of genetic diseases, deeply anxious for anything that could cure or alleviate some truly awful diseases. There are other groups who voice profound concerns about the implications of some genetic research for our treatment of animals.

Radford pointed out that it would be a mistake if people turned away from the science, because until they understand the science they cannot make sensible political decisions about it. He is right that science must inform political decisions, but it does not have to determine them. However, a frequent mistake of both the scientific community and industry is to assume that correct decision making is primarily a matter of educating the public, who will then see the problems as science sees them and so realise that their fears are groundless. This did not work for nuclear power and there is a danger of the mistake being perpetuated for genetic engineering. By looking only at the technical logic of the genetic paradigm, if this becomes seen as the new context in which to view living organisms including ourselves, many wider public concerns based in value rationality are likely to be missed. The lesson seems not to have been learned that there are many other questions than the mere logic of scientific progress which must be asked.

As indicated, one of the key questions is who benefits and who loses. Radford again observes there could be 'many happy returns for children who might otherwise die, and fairly happy returns for those who had the patents, too.'[239] This is not always a win-win situation, however. DNA technology has been big business for years and getting bigger. When the financial stakes are high, human values are apt to be sidelined and replaced by the values of the companies investing in it. Commercial logic can be seen as another form of instrumental rationality, alongside the scientific one, with both potentially in conflict with wider societal values. What then is the trade off between science, money, power and human, animal and environmental welfare?

These questions of competing rationality are the stuff of political argument. To help unravel them, two kinds of concern are now examined about genetic engineering in its social context, both of which are grounded in types of rationality. One is about control from a value rational standpoint of democratic accountability. The other is with concerns about consequences and the limits to instrumental rationality. The two questions in the end are linked.

Who Controls?

A Democratic Deficit

One of the most striking reactions to come out of the Dolly sheep cloning case was that of surprise. Despite the fact that the precursor to the discovery had been front page news a year before, until it was displaced by the Dunblane massacre, it seemed that people and politicians alike were taken aback. This was exemplified by President Clinton calling for an immediate inquiry to look at the status of human cloning, given that US law appeared to have no controls over research that might begin in the private sector. A general feeling emerged that this discovery was thrust unawares upon an unsuspecting public who felt they ought to have known it was happening. Given the publicity given to the Roslin Institute's earlier publication, this may not be entirely justified. But it points up how remote people feel from what is going on in non-human biotechnology, or from having any sense of a real say over the developments being made or the products which will eventually be offered. This is also true in the area of genetically modified foods. And with a lack of control comes the risk that eventually this could turn them against further genetic developments.

The banner of concern with the question of who controls has for some years been taken up by the publications of various groups concerned about developments in genetic engineering, such as *GenEthics News* and the Bulletin of the Genetics Forum, *The Splice of Life*. These are fairly well informed scientifically but actively seek to set the various genetic developments in a legal, ethical, political and economic context. As such they seek to exercise a critical and campaigning function aimed both at scientists and the general public.

The overall concern was illustrated in an editorial entitled 'The Democratic Deficit'.[240]

> *Many stories in this issue of* GenEthics News *illustrate the problem of trying to gain even minimal democratic control over new technology – and the partial successes that can sometimes be achieved. The main problem is that in the absence of politicians prepared to take up the issues, or a campaigning citizens' movement, decisions on genetic engineering are taken in the twilight world of experts and government advisory committees which are heavily dominated by ...genetic engineers, often representatives of industry... We are a long way from exerting democratic control even over legislation on genetic engineering and how it operates, let alone over funding decisions which determine how the technology is applied. The challenge for the 1990s is to find ways to make democracy work.*

A routine feature of the publication, under the heading 'Coming to a Field Near You' are details of the release of genetically engineered organisms, drawn from the public register of information held at the DETR. 'The aim is to let you know what may be going on in your own backyard'. This points up two problems. Certain information is publicly available, according to the strict letter of the law, but is not at all easy for the average member of the public to know of its existence at times when they might have cause to want to know. On issues with consequences as potentially far reaching as those being raised by genetic engineering, this seems a failure in communication. It is also likely to have the unintended consequence of polarising the issues, if the public's information channel is one that is mediated via organisations who are in principle opposed to certain of the developments. It is clear that if real public debate is desired, it is not sufficient to have a file open in an office in London or Edinburgh. Local registers exist, but it must be asked if there are also other ways to make the information better known.

The second outcome is that the information is selective. The constraints of commercial confidentiality may restrict what is made available to the public. Case Study 7 on BST is a case in point. There is obvious public concern to know the basis for food safety decisions being made on its behalf. Doubts have been raised from various sources over whether the scientific data considered in the early 1990s by the relevant UK veterinary and medical committees had been properly analysed by the companies concerned before it was presented to them. Whatever the rights or wrongs of this, it would appear that most of the data have remained locked away from public scrutiny in the UK, apparently at the behest of the companies concerned, because of being commercially sensitive. This suggests a failure of democratic accountability, if too great an emphasis is being placed on the values of the world of commerce, including secrecy, and too little attention is being put on wider societal values such as trust, involvement and transparency on matters of public safety. One form of rationality is being placed in an unequal power relation to the other.

Interest Groups and Society

The reference to the backyard in the previous example is a reminder that activist arguments can range from 'not in my backyard' (NIMBY) to 'not in anybody's backyard' (NIABY). Evaluations of the consequences of the environmental release of genetically engineered organisms or genetically modified food depend on one's context. To some, the NIMBY argument is inherently selfish and parochial, if it sets local concerns against the more general good of a society. Those making their local case argue that their opposition is not necessarily based on unthinking stubbornness. There may be deeper and valid concerns, relating to the wider intentions and consequences of the proposed actions. Depending on how satisfied people are with the answers, the local proposals may be supported or not.

The kinds of conflict that can arise are illustrated by Case Study 5, over the genetically engineered virus containing a scorpion toxin gene as potential pesticide against caterpillars which destroy valuable crops. Field tests were carried out, but there was considerable controversy among scientists about these tests and what they signified. These are in addition to the concerns of various interest groups like Butterfly Conservation and Friends of the Earth, and the local residents who were concerned at the proximity of the experiments to the Wytham Woods nature reserve and certain rare species of moths. The inherent problem facing experimenters in such situations is that no one can say ahead of the experiment that there is no danger to the environment. It is a risk assessment question.

If the local objections about what is going on in a particular area are generalised, they may become a universal view that such a development should not be in anybody's backyard. This point has become more complex through emergence of a whole subculture of flying pickets and organised civil disobedience, so that a variety of external agendas may be added to the local issue which, rightly or wrongly, set it in a far wider context. *GenEthics News* reports on local activities in different countries using an international network of information. Some of this carries with it the notion that local objections could and should be universalised. Under the headline Activists Destroy Transgenic Sugar Beets, it reported:

> *The German Gen-Ethic Network reports that on the night of June 12th persons unknown destroyed 2850 genetically engineered sugar beet plants which were part of a field trial. The trial, conducted by a German company KWS, was intended to test the resistance of the plants to a virus known as BNYV. In 1993, there were many public objections to the trials, and another site of the experiment was occupied but not destroyed by a group called Arche GENoah. In Holland, in previous years, trials of transgenic potatoes were wrecked by various groups with names such as 'The Seething Spuds'.*[241]

As with most activists, motivations and reasons for action may vary considerably. Some will be opposed on the ethics of conviction that transgenic work is inherently wrong. Others will be concerned about the democratic deficit, that such experiments take place without proper public discussion or accountability. Still others will be concerned about the particular consequences, which inevitably raises conflicts of interpretation. Thus it has been pointed out by scientists involved with the field trials that those who destroy the crops have done so by using environmentally damaging toxic chemicals. Indeed, such actions are if anything likely to enhance the risk of foreign genes escaping as a result of scattering vegetative and potential seed material into the surrounding environment. Since the destruction is supposedly done in the context of environmental politics, it would seem the activists do indeed have a case to answer. They might rationalise their action on the basis of risk comparisons, that the risk of scaling up field trials is worse than any risk arising from destroying the field trials.

This continuing trend of disruption raises an important question – where does this leave the democratic deficit? There is concern with commercial secrecy and activities that are carried out by experts who are not effectively accountable to the wider community. By exactly the same token, however, it must be asked who the activists are accountable to, and what actions are truly representative. There seems to be a democratic deficit both ways.

Keeping Democracy Alive in a Plural Culture

The key issue is not about simply any action, which is ultimately anarchy, but rather what democratic action is both representative and accountable? This was precisely the issue raised by Beck-Gernsheim comment with which this chapter began. How can we keep democracy alive in an area in which decisions are so fundamental that they may significantly alter the course of our future? For her part Beck-Gernsheim poses the problem but confesses that she has no patent remedy to offer. But given the complexity of the problem, the path to be taken may turn out to be something of a labyrinth. What would the outlines of a solution look like?

First, it will be grounded in some form of institutional pluralism. Democracy, after all, is based on considerations of checks and balances. There are numerous legal, economic, scientific, religious and political institutions which come into play. Each of them operates with its own ground rules within which conflicts of interest may be regulated. We need to examine how they currently interconnect in practice. In the end, it is the political sphere where matters concerning the public interest are debated and decided. The question then is how to develop a political culture in which politicians are educated to appreciate the scientific and ethical issues involved as well as, say, the political and commercial aspects. Parliamentary select committees, such as that on science and technology, can be quite effective in obtaining scientific data, but tend not to be geared to a comparable level of ethical assessment, nor to a critical examination of how the context for the questions is framed. Who will guard the guardians? This is a continuous educational task, but it is a prerequisite for informed judgements. Those who have responsibility for making such decisions should know about the range of alternatives and why they have made their judgements in the way they have. This can never be done infallibly, but it is the ground on which representative and responsible political action can be taken.

Secondly, we have to consider the role of the public. Here again, there is an educational task. It is not adequate to produce a culture either in which experts are treated as god-like, or one in which they are totally dismissed because they disagree with each other. Indeed, it has been clear from the BSE situation that people can shift from unqualified trust to general mistrust when they have reason to feel let down. Through education we can learn respect for knowledge, but through it we also become aware of how knowledge is limited and provisional. There are ways in which citizens can develop the faculty of questioning and evaluating what it is they are being told. There is no substitute for an educated public. This is why a society with democratic pretensions has to take education seriously as a long term and continuing project. This at least allows for some critical space and makes accountability more feasible. Moreover, it is out of such a climate that politicians and experts need to come in the first place. Without this, even if we accept the point about institutional pluralism, we will only be dealing with elite pluralism not democratic pluralism.

Thirdly, we have to give serious attention to the role of the media. We know that the media can trivialise and sensationalise news on the big issues of the day. Yet they can also, through high quality investigative journalism and professionally competent and specialist journalists, inform the general public of things they would not otherwise know. They also have an important role in developing critiques of what powerful interests are engaged in. This was why the press in the UK came to be described as the Fourth Estate. It is not hopelessly idealistic to recognise that things can be much better than they currently are. And if we tackle the democratic deficit in one sphere, then because of the kind of interconnections we have noted, it will inevitably have ramifications in improving the others.

RISK AND THE LIMITS OF SCIENTIFIC RATIONALITY

Risk and Scepticism

In Chapter 7, risks of environmental impact from genetic engineering were discussed. That discussion highlighted the difficulty of finding a suitable framework for prediction of inherently uncertain events, when little or no data exist about the risks, and when in some cases it may not even be possible to articulate with any accuracy what the risks are. The conclusions of the lay panel of the 1994 UK National Consensus Conference on plant biotechnology also emphasised the range of uncertainties of the impact of plant biotechnology on the environment.[242] Given the common assumption that science is concerned with prediction, these uncertainties, and the caveats that follow from them, are rather significant. Instead of exact science, we are faced with uncertainty.

This is evident when experts disagree with one another. Even when there is an expert consensus today, according to Popper's argument it will be provisional on new data appearing at some time in the future. The real paradigm shift of genetics may thus be one that is still to come – not that we have found a new language for describing biology, but that it is not as predictable as we expected. In this light, the trust of lay people can reasonably be conditioned by scepticism about the science itself. It is conditioned even more by scepticism about the idea of science driven by commercial interests, since, as we have seen, the rationality of commerce does not *a priori* tend to be altruistic in relation to wider social values. This ambiguity and complexity about genetic science cannot be wished away, nor can these concerns be dismissed as irrational.[243] But in the meantime some decisions have to be made based on inadequate data. This is the perpetual human predicament of technology.

There are several kinds of scepticism. There is scepticism as shown by the scientist, aware of the provisional nature of knowledge. There is also scepticism as shown by the lay person who asks, can we really trust scientists – since even experts can disagree or change their minds – let alone those with vested interests? Both call attention to the relationship between risk and uncertainty, and to the question of who or what we can trust. In Anthony Giddens' view, this is part of the experience of modernity:

> *In conditions of modernity for lay actors as well as for experts in specific fields, thinking in terms of risk and risk assessment is a more or less ever present exercise, of a partly imponderable character. It should be remembered that we are all lay people in respect of the vast majority of expert systems which intrude on our daily activities. The proliferation of specialisms goes together with the advance of modern institutions, and the further narrowing of specialist areas seems an inevitable upshot of technical development. The more specialisms become concentrated the smaller the field in which any given individual can claim expertise, in other areas of life he or she will be the same as everyone else. Even in fields in which experts are*

in a consensus, because of the shifting and developing nature of modern knowledge, the 'filter-back' effects on lay thoughts and practice will be ambiguous and complicated. The risk element of modernity is thus unsettling for everyone; no-one escapes.[244]

Christian thinking stresses the mutual responsibility of all players in society. Any groups whose activities and preoccupations end up in risks being taken – such as genetic scientists, biotechnology companies and genetic special interest groups – need to have their claims that these activities are 'for the common good' duly scrutinised. That indeed is one of the main aims of this book. Such scrutiny implies a rather greater degree of accountability than the fragmented picture with which Giddens leaves us. As observed in the previous section, it is a constant task for a democratic society, not just to be educated but to learn how to balance competing claims. A Christian view of life recognises responsible risk taking as normal, an element of faith in God's care and providence, but this is within the context of a relational view of life. Scepticism about experts should not mean a general retreat from taking risks, therefore, but rather learning how to be socially responsible in risk taking. At every stage in the history of technology, this has to be done in the face of less than adequate data and less than perfect human beings controlling the risks. It is a moot point whether or not the stakes are any higher with genetics than with inventing the wheel, or the ability to fabricate steel or to make controlled explosions. The common factor is the need for accountability. If the society into which contemporary inventions come is inherently more fragmented than those of our forebears, it calls for a comparably greater exercise of humility, by which all interested parties are prepared to put their activities to the test of the whole, rather than merely doing all they can to pursue their own interests.

Rationality in a Risk Society

In recent years there has been a growing awareness of uncertainty in the light of unintended consequences, which has led to extensive discussions about the notion of a risk society.[245,246,247] Beck coined the term 'risk society' as a reflection of the fact that unintended consequences are an intrinsic product of the process of industrialisation, and that they can have the effect of undermining and destroying the industrial societies that were brought into being. According to his thesis, because of the ways in which the earth's resources have been intensively and extensively exploited, nature can no longer be considered as outside society, nor society outside nature. Traditional distinctions between the natural order and the social order are much more complex. There has been a process of transformation. Threats to nature arising from culture have changed into threats to the social, economic and political order as a result of what we humans are doing with nature. This transformation has led us to a risk society – a society shaped by the risks which have become part of its fabric. Industrialisation has indeed solved some of the problems associated with pre-modern societies, but in the process it has had unintended consequences in the creation of other problems. Thus today we have the current concerns about global warming, ozone depletion, rain forests and industrial pollution. Leaving aside intrinsic concerns and opposition on the basis of the ethics of conviction, genetic engineering is one more technological example which carries with it the possibility of adverse ecological and human consequences that were unintended when the interventions took place.

Such considerations should lead to a greater awareness that discussions about risk and the distribution of risk are contextualised in different ways. Industrialists will talk in terms of cost-benefit analysis, whereas social movements will talk of worse-case scenarios and the potential for catastrophe. There are competing rationalities which are represented in different discourses, as Beck points out in discussions of damage to the ozone layer and the risks of nuclear power. The struggles and conflicts about the distribution of risk and therefore the overall architecture of how risk is defined are power struggles. The decisions that are made in concrete cases say something about the distribution of power in society and, even more generally, about the nature of globalisation. To make such decisions less opaque and more open to participation from the public sphere is itself part of an ongoing struggle to make the exercise of power more accountable and democratic. This will itself challenge notions of technological determinism and economic imperatives, which can camouflage the fact that real options are available. Part of the democratic process is to make visible what those options are and to create structures in and through which they can be examined and evaluated. Needless to say, this will not create a zero risk world! We can leave such utopias to science fiction writers.

11 FINAL REFLECTIONS

Genetic engineering is a still a young technology. It is at that exciting stage driven by the vision of the potential that could be achieved, beckoning the eager researcher and shrewd manufacturer alike. It has emerged, however, at a point where European society is more aware, and often more critical, than before of the issues raised by powerful new areas of science. Alongside the enthusiasm, much concern has also been voiced at the complex and far-reaching questions which the discoveries are raising. It is already clear that these need to be assessed alongside the technological developments, not in their wake. The attitude formerly common in both the scientific and commercial worlds, which regarded ethics as a separate and retrospective matter for society to consider, will no longer suffice. Increasingly, scientists are aware of the need to examine technology and ethics side by side.

This book has attempted to bridge the gap between these two worlds. It has done so in recognition of the plurality of views in society, reflected in the working group itself. While Christian ethics provides the backdrop for many of the reflections, it is by no means the only viewpoint examined. Indeed, though several members of the working group believe that Christianity offers uniquely appropriate insights with which to study the complexity of genetic issues, they and the rest of the group differ in their evaluations. This closing chapter does not seek, therefore, to present an agreed set of conclusions, but some final reflections on the major issues which have been explored. We identify the common threads running through the issues, to which we kept finding ourselves returning, from whichever point we started. The key themes are then summarised, and to help the reader to form his or her own views we also identify some basic questions in each area.

SOME COMMON THREADS

The following are some of the recurrent themes in the study, in order of chapter:

- the evaluation of intrinsic ethical questions as well as consequential ones;
- the mixing of genes across species;
- the use and suffering of animals, and balance of medical research against animal use;
- the balance of technological and traditional approaches to agriculture;
- the significance of transgenic material in food, and the need to label it as such;
- the balance of environmental and other benefits and relatively unpredictable hazards;
- the level of precaution implied by different risks, and when regulation can become less precautionary;
- the role of commercial drivers, especially as represented by patenting biological material;
- the tendency to an unjust sharing of costs and benefits;
- the role of social and political power structures in the direction of biotechnology;
- the importance of recognising the underlying value systems of the different players;
- the need for greater discussion and public accountability, and for a biotechnology ethics commission.

THE MORALITY OF GENETIC ENGINEERING

The popular notion that genetic engineering is playing God can be seen in two ways. Humans could be said to be called to imitate God's creative example, in which case why draw a line forbidding transgenesis rather than any other intervention in nature? On the other hand, would it be assuming a role which was not ours to have? Genes are the basic building blocks of living organisms. To manipulate them at the molecular level is to make fundamental changes with only a certain finite degree of knowledge and wisdom. Choosing to transfer genes between biologically remote species, or inducing novel behaviour in organisms goes beyond any previous human agriculture or use of nature. This should call for a humble attitude and a precautionary approach (Box 11.1).

To exclude genetic engineering as unnatural begs the question of what is meant by natural after centuries of human activity in the biosphere. Yet the expression of intuitive concern over the *level* of intervention should not be dismissed out of hand. It would be more convincing to argue that genetic engineering expresses too much overlordship than that it is inherently wrong.

BOX 11.1 QUESTIONS OF MORALITY

- Do humans have the right or the wisdom to genetically engineer other life forms?
- Is it wrong to transfer genes between unrelated species? If so, why?
- Is genetic engineering disrupting relationships in the natural order?
- Should we then continue to produce insulin by genetically modified bacteria?

The strongest grounds for intrinsic concern arise from the relational perspective found in some Christian and secular ecological approaches, concerning the disruption to patterns and relationships in the natural world that we still only partially understand. Genetic engineering is a reductionist approach to biology, whose strength is in the variance of single effects, and whose weakness is the difficulty of seeing these individual changes in relation to ecological complexity. In a more holistic view, the acceptability of a genetic modification lies less in whether it crosses over a line than in assessing its effect on the wider essence of the organism, its connections with the environment and with human social affairs. If genetic engineering is not wrong in itself, neither are all its applications universally acceptable.

RISK, REGULATIONS AND PUBLIC ACCEPTANCE

The uncertain and unforeseen aspects of genetic modification are the primary focus of concern for many people (Box 11.2). Often anxiety is expressed over whether we know enough to justify proceeding at the breakneck pace with which the technology appears to be progressing. From its early days, however, many scientists were aware of the need for precaution over manipulating such powerful basic forces. Researchers are now more confident in their understanding than in the early 1970s, but they have not yet carried the public with them in that confidence. There is thus a tension between those who wish to keep the brakes on genetic engineering developments and those who believe the time has come to relax some of the regulations on the basis that scientific data would justify doing so. This is reflected in the conflict between precautionary and reactive forms of regulation of risk, respectively based on predicting in advance what may be the risks, and calculating the risks on the basis of hazards and past mistakes.

The human imagination can fail to see consequences which in hindsight were obvious. Equally it is able to predict dangers which are not real, and

BOX 11.2 QUESTIONS OF RISK

- What do you consider to be an acceptable risk – do you smoke, drink or drive a car?
- Is genetic modification likely to lead to safer or more dangerous ways of agricultural production compared with present practices?
- How precautionary should we be about releasing genetically modified organisms into the environment?

hold back benefits unnecessarily. The view of some people that the precautionary approach is unscientific and irrational is not sustainable. On the other hand, an aversion to risk that demands minimising every conceivable hazard, no matter how tiny the probability, is equally unsustainable. Both sides are driven by powerful underlying values. In reality, less is known about some GMOs than others, justifying more or less precaution. As more are released, however, increasingly complex ecological relationships are set up, which may prove very difficult to analyse. At present there is little empirical evidence for serious problems with GMOs in the environment, but the history of all technology suggests that we can never rule out 'the one that got away' from the predictions. To many people, it is irresponsible to relax the precautionary principle too far, at this stage. It seems appropriate at present to concentrate on applications that are restricted in scale, and which confer strong human or ecological benefits.

TRANSGENIC FOOD, PRIORITIES AND LABELLING

Food is a very sensitive subject. For many people the presence of foreign genetic material in food is likely to be unacceptable – for religious or other deep-seated reasons, or due to doubts over its safety. The argument that copy genes remove any association with the species of origin of transgenic material is highly controversial. While some agreed with it, most of our group found it unconvincing. Concerns may not be universally shared, but a strong ethical case exists for mandatory labelling of all foodstuffs which have involved molecular genetic modification, for the sake of those who do so object. UK and EU regulations show serious flaws, which have led to accusations that public concerns were overruled over the introduction into Europe of unlabelled genetically modified soya and maize. Where something as sensitive as food is concerned, to market such products which cannot easily be traced and are not labelled is now widely seen as a failure to take account of public opinion. Some changes have

---

BOX 11.3 QUESTIONS OF FOOD

- Does the need to feed an expanding world population justify widespread use of genetically modified crops, to increase production efficiency?
- Is the claim to feed the world borne out by present genetic food products or by foreseeable research priorities?
- Would you buy genetically modified food if there was a likely benefit to the environment, improved nutrition, or fewer chemicals applied?

---

now occurred, but there is still widespread concern over regulation in this area (Box 11.3).

Genetic engineering in crop agriculture in general is a large and growing area. Its proponents point to emerging benefits, both human and environmental. Novel applications such as growing vaccines in plant tissue have much potential, and there is evidence that in some cases agrichemical inputs can be reduced by modifying crop plants. Environmentalists, however, point to the risks involved. They argue that there could also be increased inputs, and point to the risks involved. They put the case for a less reductionist, scientific agriculture and argue for a greater place for indigenous wisdom and for a more holistic and ecological approach. Some of us see these views as incommensurable, but others would incorporate some of the insights of ecological agriculture with biotechnological methods.

The most striking ethical claim is that genetic engineering offers the only serious hope for addressing future world food needs. Whatever the case in theory, the reality is that most applications currently coming to market are not aimed at meeting developing country food needs, but are usually western consumer products, whose primary benefit is the production efficiency of commercial enterprises. The principal factors that drive the technology have not so far been oriented to the benefits of those most in need. A substantial change is needed in research and product funding priorities if the claim is not to become devalued.

PUBLIC ACCEPTANCE

Rising public concern about issues such as genetically modified food are usually met by an emphasis on educating the public about genetic engineering (Box 11.4). Ignorance is only one reason for popular resistance, however. Neither industry nor government have taken sufficiently into account that different values can lead people to accept or reject a

---

BOX 11.4 QUESTIONS OF PUBLIC ACCEPTANCE

- How can the priorities and products of genetic engineering be more accountable to the public? What should be its main goals?
- How can the ethical and regulatory system be made to reflect public concerns better?

---

technology. Much also depends on how trustworthy the proponents of biotechnology are seen to be, and what are perceived to be their governing motivations.

Part of the democratic process is to make visible the different options for our future, and to create structures where they can be evaluated. It is clear that genetic engineering has the potential to make a major social impact, but this study has found a serious lack of public accountability over what developments we do or do not want to go ahead. The importing of unlabelled soya and maize exemplify this more general anxiety. There is considerable public concern that decisions are made in the secrecy of commercial organisations, within committees of experts, or by individual pressure groups. While assessment needs to take account of the best scientific information, an undue emphasis on the scientific, rationalist tradition tends to allow too little place for personal and societal values in decision making. On the other hand, instant trial by media is not the same thing as accountability. The role of the media in shaping public opinion on biotechnology is ambivalent, inclined by turns to exalt the potential and exaggerate the dangers. What makes good news copy may not always equate to promoting good ethical debate.

To restore public confidence in the decision making system, a standing ethical commission on non-human biotechnology is urgently needed, whose decisions and *modus operandi* are open to public scrutiny and participation. The Dutch Transgenic Animal Committee offers one potential model where expert assessment and ministerial advice are moderated by opportunities for public involvement.

ANIMAL WELFARE

The potential to cause harm and suffering is one of the main issues in the genetic engineering of animals (Box 11.5). Some of the modifications made to date are detrimental to the animal, but others are neutral or even beneficial for its welfare. Examples like the oncomouse raise considerable anxieties. Attempts at genetic modification for the direct benefit of animals

---

BOX 11.5 QUESTIONS OF ANIMAL WELFARE

• How does genetic engineering differ from selective breeding or other existing procedures in its effects on how animals are treated and looked after?
• How can time-consuming assessment of animal welfare implications keep pace with rapid development of genetic procedures?

---

have not so far proved very successful. Traits like disease resistance tend to be genetically complex, and they may often be addressed better by other means. Generally speaking, genetic modification in animals has few effects on welfare which could not also be produced by selective breeding, but the latter are also increasingly being called into question. Comparison of genetic engineering with selective breeding should not assume that the status quo is an ethically acceptable or neutral ground.

Legislative safeguards need to be maintained to protect against the welfare problems which genetic engineering can bring. High welfare standards often accompany the early stages of development but attention is necessary to ensure that, when procedures become routine, commercial pressures do not then lead to a relaxation of care. The UK is in a position to influence comparable legislation in the EU. As with other areas, better public accountability and openness are required in the procedures of committees concerned with these animal issues, if public confidence is to be ensured.

BALANCING ANIMAL ETHICS AND HUMAN BENEFITS

Three case studies of animal genetic modification for medical benefit illustrate the range of intervention in this difficult question. Few ethical or welfare problems are raised by producing pharmaceuticals in sheep's milk, but the oncomouse presents a real dilemma. Potential human benefits are sought only at the cost of serious harm to the animal. In this case the seriousness of human cancer may be sufficient reason, but there is an urgent need to review the justification for the rapidly increasing scale of use of genetically modified mice models (Box 11.6). Are realistic medical benefits being fairly adduced against the animal suffering involved; has mouse use become too routine? The dramatic increase in the overall use of model mice seems to conflict with the 'Three Rs' principle of reduction, replacement and refinement. There is a need for a culture of restraint on the use of model mice to avoid the complaint that the mice involved have been reduced to little more than material commodities.

BOX 11.6 QUESTIONS OF ANIMAL ETHICS

- Are there limits to the modification we should do to animals?
- How can we assess the stage at which potential medical benefits to humans cease to outweigh the costs to animals?
- Should we engineer animals for spare parts for humans?
- Is it wrong to clone animals?

Xenotransplantion is a novel and serious intervention in the animal kingdom, marking a change in how we relate to animals. It is different from eating them. Intuitive repulsion at the idea of having a live animal organ in one's body may convey deeper issues than mere unfamiliarity. Some of us saw it as unacceptable in principle, as a denial of the sanctity of both human and non-human life by mixing organs between species, or as an excessively instrumental use of the animal. The seriousness of intervention in the animal might, however, be justified if it offered very significant and long lasting improvement for most patients. The majority of our group took this view for the heart and kidney, given the shortfall in availability, but not as *carte blanche* for all organs. Breakthroughs in artificial hearts or repeated failures of xenotransplantation could tip the balance to unacceptable, as would a failure to address current concerns that xenotransplantation may provide a route to transmit serious animal diseases to the human population.

The therapeutic tools which transgenic animals may provide must never become taken for granted by the medical profession or its administrators, animal breeders, or the public. Mice must not become mere research catalogue items, sheep mere bioreactors, nor pigs spare part factories. An element of questioning must be essential in research or treatment. 'Is it *really* necessary to use this animal?' There are limits to how far one could go on replacing human organs, for example. Two ethical cultures both have a case – one predicated to medical treatment of serious disease as the supreme ethical goal, the other of combating the abuse of animals by humans. Neither should be regarded as absolute. A balance is implied, and it seems more likely that the current balance is weighted too far towards excessive animal use than excessive caution.

Cloning has become a *cause célèbre* but the animal implications have generally been overlooked in the speculation about human use of nuclear transfer technology. Cloning happens naturally in micro-organisms, fungi, algae and many plants, but not in mammals. For some people, cloning mammals would violate a biological distinction in reproduction. Moreover, were it to be routinely used in farm animals they would see this as bringing factory production concepts one step too far into animal husbandry. Others see the potential for improving breeding merit, but the need to maintain genetic diversity is likely to restrict uses to single step changes. It must be asked if the motivation justifies the intervention. Applications to animal diseases would be a less problematical goal than production efficiency. The main application of the technology at Roslin is, however, to improve methods of genetic manipulation of farm animals for specialist applications, for which the first signs are encouraging, to use less animals, and to open up a wider range of applications. Ethically it presents fewer problems, since cloning as such is a side effect rather than the aim, and the use is restricted. Welfare questions relating to pregnancy difficulties will however need to be resolved before it would be acceptable for general use. Extensions of nuclear transfer methods to mice will open a much greater range of cloning applications which will need careful ethical scrutiny.

PATENTING

Patenting arouses some of the greatest controversy over genetic engineering. It focuses ethical issues concerning the role of commerce in the exploitation of genetic research. Some form of intellectual property protection is accepted on the grounds of justice, protecting investment and the dissemination of commercially sensitive knowledge. *What* should be protected raises serious ethical questions (Box 11.7). Modern biotechnology has brought intense pressure from industry to regard living matter as no longer only a product of nature, but also a product of industry, and therefore patentable. Others argue that this fails to respect the normal ethical distinction between what is alive and what is not. There are strong ethical grounds that a genetic sequence or a transgenic organism are part of God's creation, for which the claim of human invention is inappropriate. A genetically modified animal may exhibit novelty in changes to one or two genes, but it remains by nature an animal. No one can claim to

---

Box 11.7 Questions of Patenting

- How can we ensure just rewards for the intellectual effort involved in genetic engineering applications?
- Is it right to claim a gentically modified animal as a patentable invention, especially if only two genes have been changed?
- Does patenting always live up to the claims of its advocates for openness and progress in research?

have invented something which is alive. Some therefore argue that either a new gene construct or the use of a transgenic organism to make a particular product could be patented, but not a naturally occurring gene or the modified organism itself. Others see no need to make such distinctions, seeing a patent as only a human device of commerce.

The primary drive is investment insurance for the large capital outlay needed to develop a biotechnological product. Exaggerated claims can be made about the benefits of patenting. At its best it can be very effective, but it can also be protracted, arduous and expensive, and may not always encourage the dissemination of knowledge. It can also tend to favour larger commercial organisations. A patent is a negative right which restricts others from marketing the invention but does not in itself entitle the holder to produce it. There are ethical implications in a patent, but patenting is not the place to decide on the acceptability of the invention in itself. It has become so by default, however, in the absence of an adequate system for public ethical assessment of biotechnological inventions. Many see a need for commissions at UK and European level, charged with looking at the ethical dimension of key patent applications, and with scope for public comment. These would have to be satisfied both to gain the patent, and to allow that its production was ethical.

NEED TO PROTECT THE INTERESTS OF LESS POWERFUL GROUPS IN SOCIETY

Many conflicts about the risk from genetic engineering reflect wider issues of the distribution of power in society (Box 11.8). One of the major concerns of this book has been the extent to which genetic engineering is being commercially driven without a proper level of assessment of the ethical priorities which should govern the technology. The winners are well catered for. By default, the situation of those who might stand to lose

---

BOX 11.8 QUESTIONS OF INTERESTS

- Should seed and breeding stock companies have the power to prevent farmers multiplying the improved and genetically modified lines in the next generation of crops or animals?
- Should developing countries be encouraged to use biotechnology to earn hard currency or feed their populations, or to use their indigenous approaches to agriculture?

is not being adequately appreciated. If this is true within the industrialised countries, it is even more the case in relation to the poorer nations of the world. There is a need to ensure that the biological resources of developing countries are not unfairly exploited and that these countries have adequate representation on international forums involving biotechnology. Adequate funding for bodies involved in research for the developing world should also be ensured.

POSTSCRIPT – REFLECTING ON THE FUTURE

Not all that is claimed for genetic engineering gets delivered. There can be a gap between expectations raised when a breakthrough is announced and actual delivery several years down the line. Few people outside the sphere of genetics appreciate such practical factors as the inefficiency of introducing transgenes into higher animals, the difficulty of manipulating genetic traits such as growth and disease resistance in animals, and that genetic procedures that are successful in tobacco, potatoes and tomatoes may not be so easy in cereals. What finally goes on the market is also apt to be limited to what is demanded and economic, which may not be the same as what is needed or was originally claimed.

It is difficult to predict how these numerous issues will work out in terms of public acceptance or otherwise of genetic engineering. The most likely areas of resistance will be over risks to environment and human health, genetically modified food and concern for animal welfare.

The risk of releasing GMOs into the environment is likely to remain a fundamental issue of concern for many people and groups. Short of a major catastrophe happening, it is unlikely that modern, technologically advanced societies, as well as some developing countries like China, will forgo the potential benefits of genetic engineering because of the hazards which GMOs may represent to the environment, to biodiversity, or to poor farmers in the developing world. There may indeed be some important environmental benefits, but as yet few tangible benefits for agriculture in developing countries are in prospect without a major re-orientation of the research priorities of the industrialised countries.

Genetically modified food has emerged as a major area of public debate. Concerns are expressed over ethical, environmental and especially health risks, over which perceptions are acutely sensitive. Genetically modified foods may ultimately take their place on supermarket shelves, but the future of this sector depends on whether companies developing such products are prepared to listen to the public. It might only take one major scare for genetically modified food to go the way of irradiated food.

In the medical sector, the acceptability of therapeutic proteins from sheep's milk, plant virus technology or xenotransplantation will all be assessed on whether they work, what their side effects are, and the extent to which the health budget can afford them. At present the prospects look good for the first two and uncertain for the third. Debates on animal disease models will, however, remain, as there are strong lobbies on both sides, and no signs of resolution. This in part reflects the different perceptions of the consequences, but also the fundamental intrinsic issues which are also present on both sides. The nuclear transfer cloning of a sheep from somatic cells met with unprecedented worldwide public attention. The technique was developed primarily as a specialist tool for more precise genetic engineering, but the signs are that it will find wider application in human disease modelling and research using animals, and perhaps in some areas of animal breeding. Some people have profound ethical concerns about the latter, quite apart from the widespread objection to any application of the methods to cloning humans.

Over time, many consequential arguments will either be resolved or shown to be justified. It is clear that intrinsic objections will, however, remain over certain aspects of genetic engineering, raised, among others, by some Christians, members of other faiths, environmentalists and groups concerned with justice for developing countries. If as a society we choose to accept the spread of genetic engineering in non-human life forms, it seems that there will need to be changes to regulatory, consultative and trading procedures, in order to address some of these concerns. The push for more biotechnology, driven by its perceived economic and job prospects, ignores these wider issues at its long term peril.

APPENDIX 1 GLOSSARY

This lists some of the more important scientific, ethical, theological and sociological terms used in the book. Our thanks are due to Dean Madden of the University of Reading for permission to include some parts from his glossary. **Bold** type is used to indicate another term in the glossary.

Abiotic Stress A challenge to a living organism caused by physical and chemical components of the environment, eg extremes of climate and water supply or mineral deficiency or excess.

Alpha–1–antitrypsin (AAT) A protein made in the liver which regulates the amount of the **enzyme** trypsin in the lung wall. Its production is impaired in the disease emphysema, resulting in damage to the lung wall, which can eventually be fatal. AAT is used as a drug to treat both this condition and cystic fibrosis.

Amino acid The chemical building blocks from which all proteins are made. There are twenty different sorts of amino acid that occur commonly in proteins.

Antisense gene A copy of a **gene** in which the linear order of the bases is reversed, making it inactive.

Back-crossing A method used in breeding, where a cross-bred plant or animal is crossed with one of the original parent strains.

Bacteriophage A natural **virus** of bacteria.

Base One of the chemical sub-units found in **DNA** and **RNA**. In DNA, the bases are adenine (A), thymine (T), cytosine (C) and guanine (G). Uracil (U) is found in place of thymine in RNA molecules. Groups of three of these bases form a code which results in one particular **amino acid** being added on to a chain of amino acids, forming a protein.

Biodiversity	The rich diversity of genetic forms in the world.
Biofuel	Alternatives to liquid fossil fuels, derived from plants.
Biosphere	The whole region of land, the oceans and atmosphere which is inhabited by living organisms.
Biotechnology	The manipulation of biological systems or processes for industrial applications.
Cash crop	A crop grown to generate income rather than for consumption, generally for developing countries to earn western currency.
Cell culture	A **cloned** population of individual cells, grown in an artificial chemical medium which enables them to flourish.
Chimaera	An organism where individual cells do not have the same genetic composition as each other. Chimaeras are produced as an intermediate stage in some methods of **genetic modification.**
Chromosome	The structure containing DNA that carries genetic information. Humans, for example, have 23 pairs of chromosomes in their body cells.
Clones (of cells)	A group of genetically identical cells.
Clones (of DNA)	Identical copies of a particular piece of **DNA.**
Clones (of a plant or animal)	Plants or animals with exactly the same **DNA** composition. In plants these are usually propagated from adult mature cells or tissues as in plant cuttings (the Greek word *klon* literally means a cutting) or bulbs.
Coat protein	As **virus coat protein.**
Complement	A group of enzymes present in blood serum responsible for causing the **hyperacute rejection** of organs transplanted between species.
Consequential ethics	An **ethical** framework where an action is deemed right or wrong because of the effects or consequences it has.
Consequentialism	An **ethical** approach where only **consequential** arguments are seen as valid and permissible.
Copy gene	A copy of a **gene** taken from a natural organism and transferred into a medium in which it can be replicated. The replica genes are then inserted into

a host organism. Thus, although the copy gene is identical to the original, its chemical components no longer come directly from the source organism but from the copying process.

Crown gall	A cancer (canker) on plant tissue.
Cultivar	A genetic **variety** of a plant **species**.
Cytogenetics	Work with plant or animal **chromosomes**.
DNA	Deoxyribonucleic acid. A biological **polymer** which comprises the heritable genetic material in most organisms. It consists of a long chain of smaller chemicals, called bases, attached to an inert backbone of alternating sugar and phosphate molecules, arranged in a helical structure.
Differentiation	The process by which cells become specialised in some function eg skin cells or muscle cells.
Ecocentric	An ethical approach which considers the effect of an action on the whole of the ecosystem.
Efficacy	The ability to produce the result intended.
Electroporation	The technique of using an electric field to transfer external **DNA** into cells.
Embryonic stem cells	**Cell cultures** derived from embryos, in which cell multiplication occurs but differentiation does not.
Erythropoietin	A protein which stimulates the production of red blood cells.
Ethics	Applying **moral** values in different situations. It may involve finding a balance between two or more opposing moral values.
Enzyme	A protein that acts as a biological catalyst.
Fatty acids	Components of oils and fats. They consist mainly of carbon, hydrogen and some oxygen.
Gene	A section of **DNA** which results in the production of a particular protein.
Genetically	Genes which tend to be inherited together and which
linked	hence give rise to traits which tend to be inherited together.
Genetic engineering	The identification and manipulation of pieces of heritable material. It allows, for example, the

production of new combinations of genetic material by transferring **DNA** into an organism in which it does not naturally occur. Also referred to as **genetic enhancement, gene manipulation, genetic modification** and **targeted genetics.**

Genetic enhancement   As **genetic engineering.**

Genetic modification   As **genetic engineering.**

Gene manipulation   As **genetic engineering.**

GMO   Genetically modified organism. An organism resulting from **genetic manipulation.**

Gene therapy   The replacement or repair of what are regarded as defective **genes** in humans.

Genome   The complete set of genetic information of an individual. It comprises all of the **DNA** contained in a single set of **chromosomes** (for example found in a sperm or egg).

Genotype   The genetic constitution of an individual organism.

Germ cells   The cells from which sperm and ova develop.

Germplasm   A collective noun for one or usually more **varieties**, **species**, etc, capable of being used in breeding or propagation programmes.

Gradualism   Where change takes place in small steps, each one justified ethically by the previous one, such that the end result, had it been achieved in a single step, would have been considered unacceptable.

Halophyte   A plant which naturally survives salty conditions.

Herbicide   A chemical which kills plants.

Heritable material   The material which passes from one generation to the next (usually **DNA** in one set of **chromosomes**).

Histocompatibility antigens   Molecules present on the surface of cells which are responsible for the rejection of transplanted organs.

Hybrid   The result of crossing between two species or varieties.

Hyperacute   The body's response to tissue which is very foreign

rejection	(eg from another species) which causes the tissue to be rejected very rapidly.
Immuno-suppressant	A chemical which prevents the body's rejection mechanism from working.
Insecticide	A chemical which kills insects.
Intra-specific crossing	Crossing within a **species.**
Inter-specific crossing	Crossing between different **species.**
Intrinsic ethics	An ethical framework where an action is deemed right or wrong in itself rather than as a result of its consequences.
Instrumental rationality	Behaviour that is logically consistent in relation to decisions about the appropriateness of given means to reach given ends (as opposed to **value rationality**).
Lactoferrin	Protein secreted in milk which is believed to assist iron uptake by the offspring or by humans drinking it.
Ligases	Enzymes which stick pieces of **DNA** or **RNA** together. They act by reforming the sugar-phosphate linkages in the backbone of the polymer.
Messenger RNA	**RNA** involved in the transfer of information which results in the **transcription** of **DNA** to form proteins.
Morals	The values which are held to be true or important by individuals and societies.
Mutagen	Something which causes **mutations**. This can be a particular chemical or certain types or radiation.
Mutagenicity	Tendency to cause genetic mutations.
Mutation	A random change in the structure of **DNA, RNA** or the **chromosome.**
Nuclear transfer	A method used in animal embryology in which the nucleus of one cell is transferred to another cell from which the nucleus has been removed. The technique has been used for example to produce **cloned** sheep and cattle.
Nucleus	The membrane-bound structure inside a plant or animal cell which contains the **chromosomes.**

Bacteria do not have a nucleus, but have 'naked' DNA as their **genome.**

Oncogene
A gene with the potential to cause cancer

Oncomouse
A mouse whose **genotype** has been modified to include an **oncogene.** The first such mouse produced at Harvard University.

Osmolyte
A compound which naturally facilitates the uptake and retention of water.

Out-crossing
A method used in breeding, in which two unrelated individuals are bred, as opposed to **back-crossing.**

Pantheistic
The belief that God resides in everything and everything in God, as opposed to God being separate from what God creates.

Peptide
Building-block of proteins, usually composed of 2–12 **amino acids.**

Pesticide
A chemical which kills one or more organisms deemed to be pests.

Phenotype
The observable characteristic of a living organism.

Phyto-hormone
A plant hormone.

Phytosanitary
Describing something involved in plant disease control or plant health.

Plasmid
A small, closed loop of double-stranded **DNA** that can multiply to very high numbers in a single bacterial cell and is used to replicate (**clone**) DNA.

Plural society
A society in which many different belief systems are found.

Polymer
A long chain molecule consisting of repetitions of the same basic small units. This complex chemical is usually formed into long chains, or into rings and other complex structures.

Promoter
A special **DNA** sequence which acts as a signal for a gene to be 'switched on' or 'off' and the relevant protein produced at the correct time of development in a given cell. Also referred to as a regulatory sequence.

Protoplast
A plant cell without its protective outer walls.

Recombinant DNA
Generally refers to **DNA** which is formed by human intervention and which contains DNA sequences

	from different origins. Naturally occurring recombination may occur in certain circumstances, particularly where bacteria are concerned.
Reductionism	The way of thinking in science in which complex systems are split into simpler entities for easier understanding. A philosophical system which extends this idea into the whole of life and decrees that things are primarily what they are at the smallest level of explanation.
Regulatory sequence	As **promoter.**
Replicon	A piece of **DNA** which has the capacity to replicate and is used in **genetic engineering** (eg a **plasmid** or **bacteriophage**).
Restriction endonuclease	An **enzyme** which cuts **DNA** at a specific place, eg where a particular arrangement of bases occurs.
RNA	Ribonucleic Acid. Material present in all living cells with roles essential for the synthesis of proteins. It is a similar biological polymer to **DNA** but, unlike DNA, RNA is usually single stranded, has the sugar ribose instead of deoxyribose and the base uracil instead of thymine.
Sentience	The ability to feel sensations such as pain.
Site of integration	The position in a **chromosome** where introduced **DNA** is chemically integrated into the host DNA through the action of **enzymes** (**ligases**).
Somaclonal variation	A genetic instability which often occurs during plant tissue culture, resulting in a **phenotypic** change.
Species	A group of organisms which are capable of interbreeding to produce viable offspring.
Stem cells	As **embryonic stem cells**
Sustainability	An elusive term, denoting an approach to personal, community, economic and industrial life in which models dominated by notions of economic growth, consumption and short-term horizons give place to ones which take equal account of the needs of future generations to flourish. It puts more emphasis on renewable resources and staying within the carrying capacity of the environment.

Symbiosis	A gathering of organisms, the existence of which is of mutual benefit to all.
Targeted genetics	As **genetic engineering.**
T-DNA	A portion of the tumour-inducing (or Ti) **plasmid DNA**. It is contained in the bacterium *Agrobacterium tumefaciens*. When it invades a plant, this part of the plasmid DNA is incorporated into the plant DNA and will normally cause the production of **crown gall**. It can also be used to introduce other genes into plants when it is normally disarmed to prevent crown gall formation.
Technicism	In philosophy, where the scientific and technological way of describing the world is used inappropriately as though it were the main determinant of our values.
Teratogenicity	The tendency to cause birth defects.
Theocentric	An ethical approach which is God-centred.
Transcription	The process of copying **DNA** into single-stranded **RNA**.
Transformed plant	A plant which has been genetically modified by **DNA** transfer.
Transgene flow	The ability of **genes** to move through the ecosystem, for example via pollen.
Transgenic	An organism altered to carry and express genes from other species.
Totipotent cells	Cells which are capable of regenerating a fully mature plant or animal.
Utilitarianism	An ethical approach which claims the ultimate good to be the greatest good of the greatest number. Actions are deemed right or wrong depending on their contribution to the general good.
Value rationality	Behaviour that is logically consistent in relation to a particular value position, as opposed to **instrumental rationality**.
Variety	A minor variant of the same **species** of plant or animal. See also **cultivar.**
Vector	The vehicle by which a **cloned DNA** sequence is carried into the cell of another organism. As birds

	can be vectors of seeds; insects, fungi or worms can be vectors of **viruses**; **plasmids, bacteriophages** or viruses can be vectors of cloned DNA.
Virus	A micro-organism with a **DNA** or **RNA** core, surrounded by protein, which can only replicate itself in the cells of a suitable host organism.
Virus coat protein	A protective protein outer layer which is normally found in **viruses**. It usually confers a rigid shape to the virus and acts to protect its genetic material. There are two typical configurations. One is based on helical cylinders, consisting of many copies of one or a few types of **coat protein**. The other is based on spherical shells with hollow interior containing **RNA** or **DNA**. Also referred to as coat protein.
Volunteer crops	Crop plants which have carried over naturally to the next year's crop without being sown, for example from potato tubers or seed left after the previous harvest.
Wild-type	The normally occurring form of a **gene**.
World view	The set of beliefs and values which explain a person's view of how things are, consciously or unconsciously.
Xenograft	An alternative word for **xenotransplant**.
Xenotransplant	An organ transplanted between different species.
Xerophyte	A plant which naturally survives in very dry conditions.

APPENDIX 2 GENETIC ENGINEERING CONCEPTS AND TECHNIQUES

THE CHEMICAL STRUCTURE OF DNA

The chemical structure of DNA is based on an extremely long polymer, the backbone of which contains alternating phosphate and sugar (5-carbon; deoxyribose) molecules. Two such ribbon-like polymers are usually wound together to form a double helix. This sugar-phosphate outer structural backbone of DNA is invariant and is common to all DNA, no matter from which organism. All the DNA 'cutting and pasting' enzymes used for genetic engineering recognise and act upon this sugar-phosphate polymer structure, usually in the double-stranded helical configuration. However, the specificity for recognition and the true genetic information in DNA is contained in the near infinite possible permutations of a linear sequence of thousands or millions of so-called bases, even though each base can have one of only four possible identities. One base is connected to each deoxyribose sugar in each DNA polymer strand, and all the bases project laterally from each of the sugar-phosphate backbones into the space between the two strands of the double helix (see Figure 1.1) like rungs in a ladder. The chemistry of the bases [adenine (A), guanine (G), thymine (T) and cytosine (C)] is not important for this book. However, the critical feature for DNA is that the bases always pair up side-by-side in the same way, like pairs of jigsaw pieces – A with T, and G with C. By this means the double helical structure ensures that the linear sequence of bases in each DNA strand is like a mirror image of the information in the other strand, and that, when the two strands separate to form two daughter double helices (before cell-division), the information content in the pair of new double helical chromosomes, one to be inherited by each cell, remains constant.

RESTRICTION ENDONUCLEASES

Restriction endonucleases are enzymes from specific bacteria which cut both sugar-phosphate backbone strands of a DNA double helix at precise points relative to a predetermined recognition sequence of bases; usually in the middle of a four or six base pair site, or one base-pair in from each end. The reason that the source bacterium's own DNA is not also cut, with fatal consequences, is that a second group of enzymes (methylases) modify the same recognition sequence wherever it occurs in the bacterial chromosome. Thus the two classes of enzymes work in concert to protect 'own' and recognise and degrade only 'foreign' DNA entering the bacterium. Curiously, most so-called 'restriction sites' in DNA are also palindromes, (ie they read the same in each direction and have a symmetrical centre, taking into account the A=T; C=G base pairing rules mentioned above). In other words, sequences like GGCC or AAATTT or AGTACT or GTCGAC are specific restriction sites in double helical DNA. Conventionally, only one DNA strand is represented, because the sequence of the other strand is predictable by the A=T, G=C rule.

As each position could be occupied by one of four possible bases, the probability that, for example, a four base (GGCC) site would occur is calculated by 1 in 4x4x4x4, or once every 256 base pairs in the linear DNA sequence. Of course, a six base target site would occur less frequently, in fact, only once every 4096 bases. Four thousand base pairs (4 kilobase pairs or 4 kbp) is a manageable, and roughly gene-sized fragment of DNA. Fragments can now be prepared reproducibly and homogeneously from chromosomes because the ends are, by definition, fixed and determined by the restriction endonuclease recognition sequence. If two DNA fragments from different organisms are prepared using the same restriction enzyme they will have a compatible single-stranded DNA overhang, and can spontaneously base pair with each other (see Figure 1.2).

OTHER METHODS OF TRANSFERRING GENES TO MICRO-ORGANISMS

Efficient uptake of plasmid DNA by bacteria usually requires the formation of a DNA precipitate using a calcium salt. It is believed that the DNA-calcium complex lands on and penetrates the bacterial cell wall and membrane at sites which only exist in so-called 'competent' cells. Alternatively, a mixture of competent bacterial cells (many millions) and the genetically engineered plasmid DNA (sub-microgram amounts), in a suitable liquid, are subjected to a sudden high voltage electric field. The process is called electroporation and, as the name suggests, it is believed that the negatively charged DNA accelerates suddenly toward the anode (+) and is effectively fired into any bacterial cell that gets in the way. The sudden electric shock may also punch holes in (porate) the bacterial cell outer membranes through which some of the liquid medium (and the engineered DNA) then enters. Whatever actually happens, the end result is that a minute amount of the genetically engineered plasmid DNA enters a sub-population of the bacterial cells. Usually only one plasmid DNA molecule enters a particular cell, where it can then multiply to many hundreds or thousands of copies and, in so doing, clones the foreign gene.

USING *AGROBACTERIUM TUMEFACIENS* TO GENETICALLY MODIFY PLANTS

In the wild, the bacterium *A. tumefaciens* induces cankers or crown galls by producing plant hormones which promote uncontrolled cell growth. The bacterium also subverts the normal metabolism of infected plant cells to produce unusual compounds called opines which only the bacterium can use as a source of energy, nitrogen and carbon. To do this, the bacterium transfers and integrates a piece of its own DNA (part of the so-called tumour-inducing or Ti plasmid DNA, T-DNA) into the chromosomal DNA of the invaded plant cell. In order to exploit the T-DNA transfer and integration activities of the bacterium without the parasitic and patho-genic features of its lifestyle, several 'disarmed' strains of the bacterium were made. Some of the virulence genes responsible for canker formation were identified and removed, and those Ti-plasmid genes which coded for proteins involved in transferring and integrating the T-DNA part of the Ti-plasmid were left behind, together with inverted repeat sequences (25 base pairs each) which flank the T-DNA portion. The T-DNA region was given a number of restriction enzyme sites where researchers could insert one or more foreign genes. Thus a system was created to move foreign DNA molecules into plant chromosomes.

OTHER METHODS FOR INTRODUCING GENES INTO PLANT CELLS

These have involved violent mixing and shaking with needle-sharp fibres of silicon coated in foreign DNA; calcium salt-DNA precipitates or electroporation (as for bacteria) using wall-less plant cells (protoplasts) made by digesting away the thick outer fibrous cellulose with fungal enzymes. This works quite efficiently especially for crops which cannot be host to *Agrobacterium.* Since 1987, a particle gun which uses helium gas or, previously, a blank gun cartridge to propel very small gold or tungsten particles coated in DNA into plant cells has been widely used to transfer foreign genes into a wide range of plant species, beyond the host range of *A. tumefaciens.*

AN EXAMPLE OF TISSUE CULTURE AND REGENERATION

A typical example would involve cut pieces of a tobacco leaf placed in a special medium containing *A. tumefaciens* with a disarmed Ti-plasmid containing foreign gene X and selectable marker gene *kan^r*. The leaf pieces are then washed and placed on sterile nutrient agar in a Petri dish in the presence of the antibiotic kanamycin. Untransformed cells go brown and die because of the active antibiotic. Surviving cells form small clusters of cells, and small shoots can be produced from these when the correct phyto-hormone is added to the agar. These shoots grow in the presence of the antibiotic and become green in the light. Different hormones are added to the agar which then prompt rootlet formation. The tiny plantlet is eventually transferred to fresh agar in a small transparent, sealed pot (mini incubator) where it grows. It is finally transferred to sterile compost in a flowerpot in a special containment glasshouse, where it is kept under UK or equivalent national GMO regulations.

Subsequent experiments on DNA extracted from small samples of the plant then show how many copies of foreign gene X it has, how active they are in making messenger RNA, how much protein X is being made and, most importantly, if the plant has inherited the new and desirable characteristic encoded by gene X. Many tens or hundreds of separately transformed plant lines must be made and tested in this way to select those with optimal expression of the desired trait without discernible abnormalities of growth or in traits of agronomic value.

APPENDIX 3 FRAMEWORKS FOR MAKING ETHICAL ASSESSMENTS

This group has not used any one single framework for evaluating ethical issues. As highlighted in Chapter 3, much of the discussion has used biblical principles, environmental ethics or risk-benefit assessments and the social dimension. Additionally, four methods which are presented below, outline ways which different members of the group have found helpful to guide them in thinking over the issues – the flow diagram, graphical representation, choices between criteria, and societal groupings. Each of these has its problems, and some will appeal to particular groups more than others, but readers may find help from them in their own reflections on these issues.

FLOW DIAGRAM

One approach is to follow a system of logic, in the form of a kind of event tree or flow diagram of questions on genetic engineering. This could assess genetic engineering in terms of a progression of levels of human interaction with the natural world, starting with the most basic interaction, and becoming progressively more complex and controversial. In fact this reflects the logic behind the order in which the case studies have been presented. Different people, it is argued, will find objections, and so stop at different points in the process.

The value lies in clarifying the implications of certain positions, appealing particularly to a certain type of scientific logic. Its weakness is that this strictly linear progression of logic cannot encompass the entire ethical assessment. In particular, a linear model does not by nature express sideways connections and complex interactions. It is possible to disagree at one stage but then to agree to something which occurs at a later stage in the flow, because the criteria which underlie the logic have ignored other factors by which something may be judged acceptable or not. Figure A.1 explores the acceptability of genetic engineering in progressively more complex organisms, starting with microbes and ending

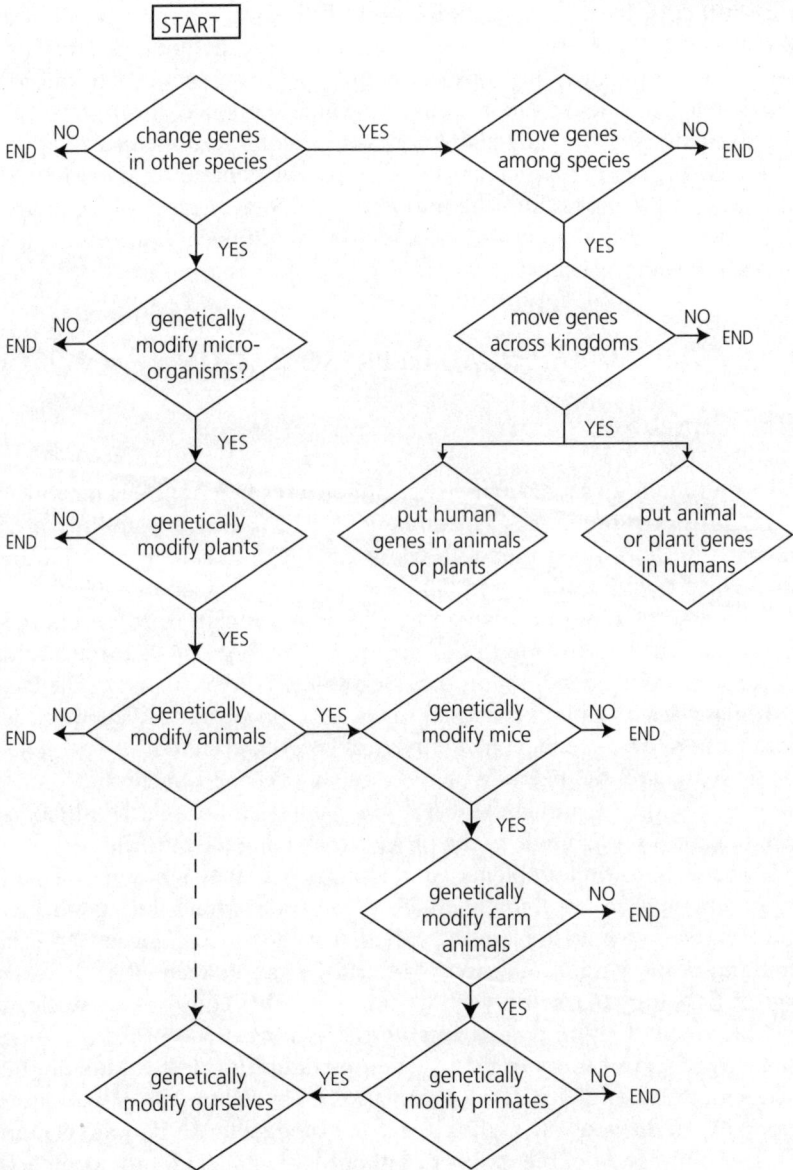

Figure A.1 *Ethical Flow Diagram – Genetic Engineering of Different Species*

with humans. It happens, however, that biological complexity is not a very useful criterion, since the crucial ethical factors do not correlate with levels of complexity. Thus risk of escape is a strong factor with micro-organisms but is less so in animals, for which sentience in turn features highly. Things which affect dignity and personhood are crucial with humans. Figure A.2 looks at the example of genetic engineering in animals, and demonstrates the importance of the contribution of different factors. It also identifies where issues do not fit linear logic, and helps locate the source of complexity.

GRAPHICAL REPRESENTATION

Bar Chart and Scores

A second method is a diagrammatic approach using a bar chart on which a range of consequences is evaluated for ethical acceptability, on a notional scale of one to nine, in which a score of 5 is neutral. A purely illustrative example is given in Figure A.3, of what someone might judge for the case of xenotransplantation. This is not meant to represent the particular view of anyone in our group, nor to suggest that one could make a rigorous quantification of ethical values. It reflects rather the fact that many scientists find it helpful to visualise problems in diagrammatic form, and to evaluate them in a numerical fashion, even if the judgements being made are, by their very nature, qualitative ones. Since there is no way to measure '3' and know that it was neither '2' nor '4', the numbers can only be a broad guide in the process of making judgements.

There are several problems with this approach. It assumes that a visual representation has meaning in ethics, which some would find untenable. Dangers can also arise from the fact that it looks scientific, when it is nothing of the kind. An illusion of objectivity can be created in the very act of drawing charts and writing numbers. This could easily mislead people (including the participants) into assuming it was based on some objective form of measurement, when by definition it lies outside the scope of the scientific method. It can also be abused by others who were not party to the assessment, but who come along afterwards and take the surface impression of the numbers without having a feel for the complexity and guesswork involved. The method also assumes that comparisons can be made, which would not apply when comparing a consequential view with a deeply held inherent ethical position.

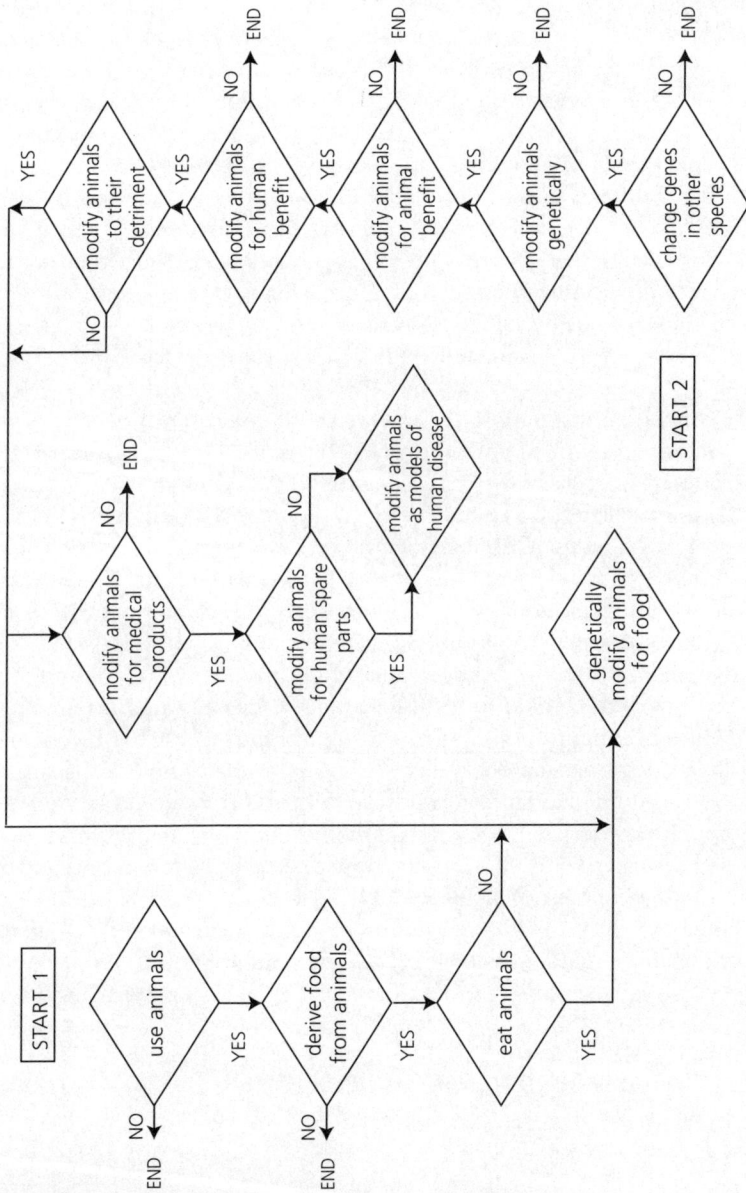

Figure A.2 *Ethical Flow Diagram – Genetic Engineering in Animals*

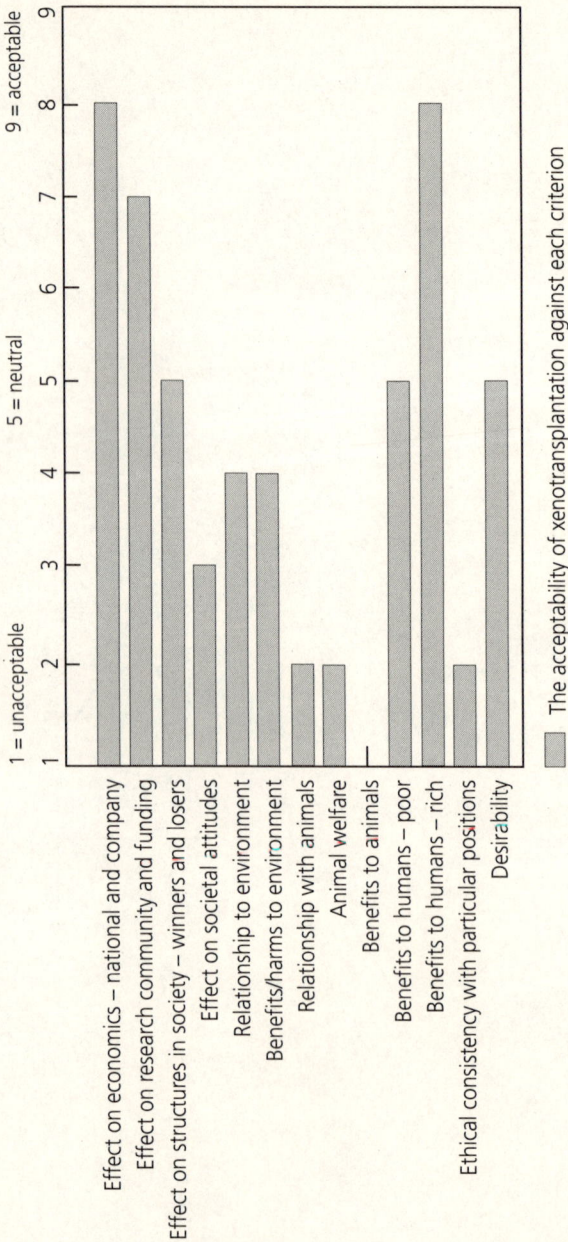

Figure A.3 *Representing and Comparing Different Ethical Considerations by Means of a Bar Chart*

Mepham's Grid

A somewhat related method was suggested previously by Mepham,[223] in which a foursquare grid is constructed, and a particular ethical value is assigned to each axis – for example greater or lesser harm along one axis, and greater or lesser justice and freedom along the other. For a given transgenic animal application, one could draw different circles from the point of view of the animal, of consumers, of biotechnology companies, or of farmers. Overlaying the circles would give an overall view of where an ethical balance might lie. Again, such a grid could be made of animal harm and human benefit, and each member of a group of people could be invited to draw a circle which represents his or her personal evaluation, and by overlaying the circles a general impression of the weighting of views within that group obtained. Obviously this method has problems of the individual weighting of the different circles, equivalent to sizing the bars in the bar chart, but it can be a schematic guide to compare what are the major effects seen from different viewpoints. Another danger is in reducing ethics to the average pattern of a population, which again tends to devalue the weight of inherent ethical views.

VALUE CRITERIA

Another technique is to ask what are the key underlying values and principles about life which relate to these questions. A list of these can be seen as the range of possible criteria against which ethical judgements are made in this area. Each person attaches a greater or lesser importance on each criterion, and so shapes a portfolio of the main motivations which drive their ethical judgements, and which act as a grid through which he or she sees the issues. If medical benefit to human beings is seen as the greatest good, that will lead to a certain ethical bias; if it is the suffering of animals, that will favour a different set of conclusions; if justice issues relating to developing countries are seen to be of paramount importance, another grid is overlaid on the issues, and so on. It is up to each person to put them into some hierarchy of importance.

The value of this approach is that it provides a tool for recognising views other than one's own, based on different relative judgements on certain factors. What it does not do is give a means for resolving the conflicts between values, either for the individual, or more especially a society, nor does it take account of the various societal influences which shape, reinforce or challenge these values. This book has set out to explore the interaction of differing perspectives such as these, and offers insights which come from a multifaceted examination. The following list gives some of the more important motivations and values which can affect attitudes towards genetic engineering.

Economic and Political

- National economic growth;
- economics emphasising environmental sustainability and community;
- production efficiency;
- creating more jobs;
- reporting back to shareholders increased profit margins and market share;
- the potential of a policy to gain or lose political support.

Societal

- National or institutional scientific excellence;
- NIMBY – not in my backyard, NIABY – not in anybody's backyard;
- scientific progress: the continual, evolving improvement of humankind through technology;
- scepticism of science – absence of trust, based on past failed expectations;
- promotion of the use of indigenous knowledge and community values;
- aversion to what is perceived as unnatural human intervention.

Human and Environmental

- Medical benefits to humans;
- maintaining or increasing world food supplies;
- justice particularly with respect to developing countries;
- reducing the environmental damage from existing agricultural methods;
- minimising any adverse health or safety impacts of new biotechnology;
- harnessing the potential for human benefit from animal and plant processes;
- reducing or eliminating suffering caused to animals by humans.

SOCIETAL GROUPINGS

A fourth approach is a more sociological one. It examines the issues in terms of the dominant values held by particular, identifiable groups in society. During this study many such groups have been identified, and their values examined. They include the genetic, agricultural and medical research communities, the biotechnology industry, environmental NGOs, government, the EU, patent lawyers, the subculture of the news and documentary media, conservative or liberal views within the Church, other faith communities, supermarket chains, consumers, animal rights organisations, patient support groups, the World Bank, developing countries, and so on. An individual or group can seek to identify amongst all these players the group whose priorities best reflect those most precious to themselves. They can formulate their own ethical approach around the values of their identified group and its aims. This is a more collective, even tribal, way of ethics, compared with the more individualist approach of the other three methods.

At least one of the group would see the question of agricultural genetic engineering being expressed most crucially in the context of a conflict between commercial interests of industry and government, allied to science, and the viewpoints represented by environmentalists and some traditional societies. Not all the members of the working group would wish to formalise things as strictly as this or that representative struggle, but this illustrates the importance of identifying these sorts of societal groupings and the collective ethical values and philosophical assumptions which they hold. Reference to the views of such groups abound throughout the report. On occasions they have been of seminal influence in our examination of the issues. The societal dimension to ethics has not been greatly addressed in other studies of the subject, but has been one the most fundamental and rewarding factors in the work of our group.

APPENDIX 4 SOCIETY, RELIGION AND TECHNOLOGY PROJECT

The Society, Religion and Technology Project (SRT) of the Church of Scotland has since 1970 been opening questions of technology and social change to Christian ethical examination. It seeks to engage those involved in technology with the ethical and social implications of their work, through expert working groups. This book is the product of one such group. It also addresses government, industry, the media and the general public, with the aim of fostering a balanced and informed understanding of the issues which confront society, and to act as a forum for interaction between experts and lay people. It acts as an advisor to the Church of Scotland on a range of technological and environmental issues, and represents the church on numerous ecumenical working groups and networks in the UK and across Europe. The Project works in many different ways:

- through expert study groups;
- producing reports, information sheets and other publications;
- in the media and through SRT's internet website;
- contributing to government and European consultations;
- public lectures and debates, and especially the Edinburgh International Science Festival;
- informing and assisting the churches to interpret contemporary technological issues.

In addition to biotechnology, its work and publications range over a wide variety of issues including the environment and sustainable development, climate change, energy policy, transport, technology risk and information technology. It also promotes debate on the issue of God and science.

The Project has a full-time scientific Director, Dr Donald Bruce. Its office in John Knox House, Edinburgh, has a small library of relevant books, papers and journals. It produces a regular newsletter, available on request. It is supported by the SRT Trust and by individual SRT Associates.

If you wish to respond to any of the issues raised in this book, or if you would like to know more about our work, please write, phone, fax or email us at:

Society, Religion and Technology Project
John Knox House, 45 High Street, Edinburgh EH1 1SR, Scotland
Tel: +44 (0)131 556 2953, Fax: +44 (0)131 556 7478
srtscot@dial.pipex.com
http://webzone1.co.uk/www/srtproject/srtpage3.htm

NOTES AND REFERENCES

INTRODUCTION

1   Carson R, (1963) *Silent Spring*, Houghton Mifflin: Boston
2   Hillman, J R and Wilson, T M A (1995) 'Plant Biotechnology: Delivering the Promise', Paper presented to the Parliamentary and Scientific Committee at the Royal Society of Edinburgh, 7 April

CHAPTER 1 EXPLAINING GENETIC ENGINEERING AND ITS USES

3   Watson, J D and Crick, F H C (1953) *Nature*, vol 171, p 737
4   Straughan, R and Reiss, M (1996) 'Ethics, Morality and Crop Biotechnology', pp 20–21, Biotechnology and Biological Sciences Research Council: Swindon
5   Straughan and Reiss (1996) *ibid*
6   Campbell, K H S, McWhir, J, Ritchie, W A and Wilmut, I (1996) *Nature*, vol 380, pp 64–66
7   Schnieke, A E, Kind, A J, Ritchie, W A, Mycock, K, Scott, A R, Ritchie, M, Wilmut, I, Colman, A and Campbell, K H S (1997) *Science*, vol 278, pp 2130–2133
8   Advisory Committee on Novel Foods and Processes (1997) 'ACNFP Report on Products Dervied from Insect-Resistant Genetically Modifed Maize', in: *Advisory Committee on Novel Foods and Processes Annual Report 1996*, p 144, MAFF Publications, London
9   Smith, C, Meuwissen, T H E and Gibson, J P (1987) 'On the Use of Transgenes in Livestock Improvement', *Animal Breeding Abstracts*, vol 55, no 1, pp 1–10
10 Pursel, V G, Pinkert, C A, Miller, K F, Bolt, D J, Campbell, R G, Palmiter, R D, Brinster, R L and Hammer, R E (1989) 'Genetic Engineering of Livestock', *Science*, vol 244, pp 1281–1287
11 Haley, C S (1995) 'Livestock QTLs: Bringing Home the Bacon?' *Trends in Genetics*, vol 11, no 12, pp 488–492
12 Powers, D A, Gongalez-Villasenor, L I, Zhang, P, Chen, T T and Dunham, R A (1991) 'Studies in Transgenic Fish' in: First, N L and Haseltine, F P (eds) *Transgenic Animals*, pp 307–325, Butterworth Heinemann: London
13 Dunnock, S, Su, H Y, Jay, N P and Bulloch, B W (1996) 'Improved Wool Production in Transgenic Sheep Expressing Insulin Like Growth Factor-1', *Biotechnology*, vol 14, pp 185–188

14 Cremers, H C (1993) 'Transgenic Bull to Sow Wild Oats', *New Scientist*, 9
    January
15 Bialy, H (1991) *Biotechnology*, 9 September, pp 786–788
16 James, C (1997) 'Global Status of Transgenic Crops in 1997', *ISAAA Briefs*,
    no 5, International Service for the Acquisition of Agri-biotech Applications
17 Fox, J L (1991) *Biotechnology*, 9 December, p 1319
18 Woolliams, J (1998) 'Uses of Cloning in Farm Animal Production', *Roslin
    Institute Annual Report 1996–97*, pp 24–25

CHAPTER 2 CASE STUDIES

19 Amin-Hanjani, S, Meikle, A, Glover, L A, Prosser, J I and Killman, K (1993)
    'Plasmid and Chromosomally Encoded Luminescence Make System for
    Detection of *Pseudomonas fluorescence* in Soil', *Molecular Ecology*, no 2,
    pp 47–55
20 de la Fuente, J M, et al (1997) 'Aluminum Tolerance in Transgenic Plants by
    Alteration of Citrate Synthesis', *Science*, vol 276, pp 1566–1568
21 Barinaga, M (1997) 'Making Plants Aluminum Tolerant', *Science*, vol 276,
    p 1497
22 Davies, S (1996) 'Soiled by Irrigation', *Guardian On-Line*, 18 April, p 10
23 Walker, K C (1995) 'Brassicas and Alternative Crops for Scotland. Industrial
    Oils for Oilseed Rape', *Proceedings of the Scottish Society for Crop Research*,
    no 10, pp 28–31
24 Walker, K C (1995) 'Oilseeds: The Next Generation', *Agro-food Industry, Hi
    Tech*, no 6, pp 37–39
25 Chapman, S N and Wilson, T M A (1997) 'Plant Viruses as Pharm-
    Implements', *Chemistry and Industry*, 21 July, pp 550–554
26 Santa Guz, S, Chapman, S N, Robertson, A G, Roberts, I M, Prior, D A M and
    Oparka, K J (1996) 'Assembly and Movement of a Plant Virus Carrying a
    Green Fluorescent Overcoat Protein', *Proceedings of the National Academy
    of Sciences of the USA*, vol 83, pp 6286–6290
27 *Nature Medicine* (1998) 'Fields of Pharmaceuticals', vol 4, no 5, May, p 535
28 Levidow, L (1995) 'Safely Testing Safety? The Oxford Baculovirus
    Controversy', *BioScience*, vol 45, no 8, pp 545–551
29 Miller, S K (1994) 'Genetic First Upsets Food Lobby', *New Scientist*, 28 May, p 6
30 Fox, J L (1994) 'FDA Nears Approval of Calgene's Flavr Savr™',
    Bio/Technology, vol 12, p 439
31 Redenbaugh, K and Hiatt, W (1993) 'Field Trials and Risk Evaluation of
    Tomatoes Genetically Engineered for Enhanced Firmness and Shelf Life',
    Acta Horticulturae, vol 336, pp 133–146
32 Bialy (1991) *op cit*, Note 15
33 Cozzi, E and White, D J G (1995) 'The Generation of Transgenic Pigs as
    Potential Organ Donors for Humans', *Nature Medicine*, vol 1, pp 964–966
34 Nuffield Council on Bioethics (1996) 'Animal-to-Human Transplants: The
    Ethics of Xenotransplantation', Nuffield Council on Bioethics: London

35 Department of Health (1996) 'Animal Tissue into Humans: A Report by the Advisory Group on the Ethics of Xenotransplantation', Stationery Office: Norwich

36 Porteous, D J and Dorin, J R (1993) *Tibtech*, vol 11, pp 173–181

37 *New Scientist*, 26 June 1993, p 4

38 US Patent No 4.736,866, issued 12 April, 1988

39 European Patent No 0 169 672, issued 13 May 1992

40 Lewontin, R C (1993) *The Doctrine of DNA*, Penguin: London

41 Wilmut, I, Schnieke, A E, McWhir, J, Kind, A J and Campbell K H S (1997) *Nature*, vol 385, pp 810–813

42 Campbell et al (1996) *op cit*, Note 6

43 Schnieke et al (1997) *op cit*, Note 7

44 Woolliams (1998) *op cit*, Note 18

CHAPTER 3 ETHICS UNDER THE MICROSCOPE

45 Ministry of Agriculture, Fisheries and Food (1995) *Report of the Committee to consider the Ethical Implications of Emerging Technologies in the Breeding of Farm Animals*, (Banner Committee report), HMSO: London

46 Poole, M (1995) *Beliefs and Values in Science Education*, Open University: Buckingham

47 Genesis 11: 1–9

48 Colossians 1: 15–17; Hebrews 1: 2–3; John 1: 1–3

49 Psalm 145: 15–16

50 Genesis 1: 26–28

51 Page, R (1986) '"The Earth is the Lord's": Responsible Land Use in a Religious Perspective' in: *While the Earth Endures*, SRT Project: Edinburgh

52 Page, R (1996) *God and the Web of Creation*, SCM: London

53 Northcott, M S (1996) *The Environment and Christian Ethics*, Cambridge University Press: Cambridge

54 Genesis 2:15

55 Palin, D (1994) 'Theological Reflections on God, Creation and Genetic Engineering', Methodist Church Conference *All Genes Bright and Beautiful*, Luton Industrial College, 13–15 May

56 Schroten, E (1992) 'The Philosophical Basis of Ethical Issues in Genetic Engineering', Methodist Church Conference, *Theological and Moral Concerns Involved in Genetic Manipulation*, Luton Industrial College, 19–21 June

57 Ponting, C (1991) *A Green History of the World*, Sinclair Stevenson: London

58 Genesis 1: 26–28

59 Genesis 2: 15

60 Matthew 6: 28–30

61 Genesis 1: 20–24

62 For example King David and King Solomon in 1 Kings 1: 38–39

63 As for example argued in Hartman T and Williams R (1993) 'The Ethics of Species Manipulation', *Science and Christian Belief*, vol 5, pp 117–138

64 Leviticus 19: 19
65 Macer, D (1990) 'Genetic Engineering in 1990', *Science and Christian Belief*, vol 2, pp 25–40
66 See for example in Leviticus 25
67 Beck, U (1986) *Risk Society*, Sage: London
68 Midgley, M (1992) *Science as Salvation*, Routledge: London
69 Schuurman, E (1980) *Technology and the Future*, Wedge Publishing: Toronto
70 Oliver Wendell Holmes, quoted in Cooke, A (1974) *Alistair Cooke's America*, BBC: London, p 389

CHAPTER 4 GENETIC ENGINEERING AND ANIMAL WELFARE

71 Raichon, C, Demeyer, D, Blockhuis, H J and Wierenga, H K (1993) 'An Introduction to the Workshop', *Livestock Production Science*, vol 36, pp 1–3 [The workshop papers on 'Biotechnology and Animal Welfare' form the rest of issue vol 36 (1)]
72 Banner Committee report (1995) *op cit*, Note 45
73 Appleby, M C (1998) 'Genetic Engineering, Welfare and Accountability', *Journal of Applied Animal Welfare Science*, vol 1, pp 255–273
74 Pursel et al (1989) *op cit*, Note 10
75 Robinson, J J and McEvoy, T G (1993) 'Biotechnology: The Possibilities' *Animal Production*, vol 57, pp 335–352. Quotation from p 348
76 Loew, F M (1994) 'Beyond Transgenics: Ethics and Values', *British Veterinary Journal*, vol 150, pp 3–5
77 Pursel et al (1989) *op cit*, Note 10
78 McNeish, J D, Scott W J and Potter, S S (jnr) (1988) 'Legless, a Novel Mutation found in PHT1–1 Transgenic Mice', *Science*, vol 241, pp 837–9
79 Loew (1994) *op cit*, Note 76
80 Cameron, E R, Harvey, M J A and Onions, E E (1994) 'Transgenic Science', *British Veterinary Journal*, vol 150, pp 9–24
81 Poole, T B (1995) 'Welfare Considerations with Regard to Transgenic Animals', *Animal Welfare*, vol 4, pp 81–85
82 Mepham, T B (1993) 'Approaches to the Ethical Evaluation of Animal Biotechnologies', *Animal Production*, vol 57, pp 353–359
83 Broom, D B (1993) 'Assessing the Welfare of Modified or Treated Animals' *Livestock Production Science*, vol 36, pp 39–54
84 Hughes, B O, Hughes, G S, Waddington, D and Appleby, M C (1996) 'Behavioural Comparison of Transgenic and Control Sheep: Movement Order, Behaviour on Pasture and in Covered Pens', *Animal Science*, vol 63, pp 91–101
85 Appleby, M C, Hughes, B O and Savory, C J (1994) 'Current State of Poultry Welfare: Progress, Problems and Strategies', *British Poultry Science*, vol 35, pp 467–475
86 Mepham (1993) *op cit*, Note 82

87 Banner Committee report (1995) *op cit*, p 3, Note 45
88 Coghlan, A. (1995) 'Altered Animals Need Watchdog to Protect Them', *New Scientist*, 11 March, p12
89 Brom, F W A and Schroten, E (1993) 'Ethical Questions Around Animal Biotechnology: the Dutch Approach', *Livestock Production Science*, vol 36, pp 99–107

CHAPTER 5 ANIMAL ETHICS AND HUMAN BENEFIT

90 Sandøe, P and Holtug, N (1993) 'Transgenic Animals: Which Worries are Ethically Significant?', *Livestock Production Science*, vol 36, pp 113–116
91 Northcott (1996) *op cit*, Note 53
92 Banner Committee report (1995) *op cit*, Note 45
93 Reiss, M and Straughan, R (1996) *Improving Nature*, Cambridge University Press: Cambridge
94 Regan, T (1988) *The Case for Animal Rights*, Routledge: London
95 Midgely, M (1983) *Animals and Why They Matter. A Journey Around the Species Barrier*, Penguin Books: Middlesex
96 Linzey, A (1987) *Christianity and the Rights of Animals*, SPCK: London
97 Linzey, A (1991) 'The Moral Priority of the Weak. The Theological Basis of Animal Liberation', pp 25–42 in: *The Animal Kingdom and the Kingdom of God*, CTPI Occasional Paper No 26, Centre for Theology and Public Issues, Edinburgh
98 Genesis 1:26
99 Barclay, O (1992) *Animal Rights: a Critique, Science and Christian Belief*, vol 4, no 1, p 49
100 Singer, P (1990) *Animal Liberation*, second edition, Cape: London
101 Rolston, H (1988) *Environmental Ethics: Duties to and Values in the Natural Environment*, Temple University Press: Philadelphia
102 Fox, Michael (1990) 'Transgenic Animals: Ethical and Animal Welfare Concerns', pp 31–51 in: Wheale, P and McNally, R (eds) *The BioRevolution: Cornucopia or Pandora's Box?* Pluto Press: Winchester
103 Verhoog, H (1992) The Concept of Intrinsic Value and Transgenic Animals, *Journal of Agricultural and Environmental Ethics*, vol 5, no 2, pp 147–160
104 Holland, A (1990) 'The Biotic Community: a Philosophical Critique of Genetic Engineering', pp 166–174 in: Wheale, P and McNally, R (eds) *The BioRevolution: Cornucopia or Pandora's Box?* Pluto Press: Winchester
105 Northcott (1996) *op cit*, Note 53
106 Barclay (1992) *op cit*, Note 99
107 Genesis 9 : 1–7
108 Genesis 1 : 26–27
109 Finnis, J (1980) *Natural Law and Natural Rights*, Clarendon Press: Oxford
110 Nuffield Council on Bioethics (1996) *op cit*, Note 34
111 Matthew 10: 29–31
112 Banner Committee report (1995) *op cit*, Note 45
113 Wilmut et al (1997) *op cit*, Note 41

114 Church of Scotland (May 1997) *Supplementary Reports to the General Assembly and Deliverances of the General Assembly 1997*, 'Board of National Mission Supplementary report on the Cloning of Animals and Humans', p 36/22, and 'Board of National Mission Deliverance 35', p 16 of the Deliverances

115 Church of Scotland (May 1997) *ibid*

116 Dominko, T et al (1998) 'Bovine Oocytes as a Universal Recipient Cytoplasm in Mammalian Nuclear Transfer', *Theriogenology*, vol 49, p 385; Mikalipova, M et al (1998) 'Bovine Oocytes Cytoplasm Reprogrammes Thematic Cell Nuclei for Various Mammalian Species', *Theriogenology*, vol 49, p 389

117 Nuffield Council on Bioethics (1996) *op cit*, Note 34

118 Lechler, R (1998) 'Tackling the Challenges of Xenotransplantation', *Proceedings of International Conference on The Cloning Dilemma*, Dubai, 4–5 April, in press

119 British Union for the Abolition of Vivisection (BUAV) and Compassion in World Farming (CIWF) (1993) 'Opposition Under Part V of the European Patent Convention to the Grant of a Patent for the Oncomouse to the President and Fellows of Harvard College', 8 January, Case Number T19/90–3.3.2

120 Zutphen, L F M and van der Meer, M (eds) (1997) *Welfare Aspects of Transgenic Animals: Proceedings EC Workshop of October 30, 1995*, Springer: Berlin Heidelberg

121 Weiss, R A (1998) 'Transgenic Pigs and Virus Adaptation', *Nature*, 391, pp 327–8

122 Butler, D (1998) 'Special Briefing on Xenotransplantation', *Nature*, 391, issue 6665

123 *Animal Tissue into Humans: a report by the Advisory Group on the Ethics of Xenotransplantation*, UK Dept. of Health, HMSO, 1997

124 Butler (1998) *op cit*, Note 122

125 Banner Committee report (1995) *op cit*, p 8, Note 45

126 Nuffield Council on Bioethics (1996) *op cit*, Note 34

CHAPTER 6 TRANSGENIC FOOD

127 Smith, J E (1996) *European Journal of Genetics in Society*, vol 2, no 1, pp 15–24

128 Hillman and Wilson (1995) *op cit*, Note 2

129 FAO (1995) *Dimensions of Need: An Atlas of Food and Agriculture*, FAO: Rome

130 Advisory Committee on Novel Foods and Processes (1997) *op cit*, Note 8

131 Burke, D (1997) 'The Recent Excitement over Genetically Modified Foods', Lecture to Scottish Society for Crop Research, SCRI: Invergowrie, June

132 James (1997) *op cit*, Note 16

133 Ministry of Agriculture, Fisheries and Food (1993) *Report of the Committee on the Ethics of Genetic Modification and Food Use*, (Polkinghorne Committee report), HMSO: London

134 For example 'You must be careful to distinguish between what is ritually clean and unclean between animals that may be eaten and those that may not' (Leviticus 11: 47)

135 See especially in Deuteronomy 22

136 Martin, S and Tait, J E (1993) *Release of Genetically Modified Organisms: Public Attitudes and Understanding*, Open University: Buckingham

137 Polkinghorne Committee report (1993) *op cit*, Chapter 3, Note 133

138 Ministry of Agriculture, Fisheries and Food (1995) *Genetic Modification and Food, a Guide from the Food Safety Directorate*, MAFF: London

139 Burke (1997) *op cit*, Note 131

140 Supposing it were ever possible to perform gross genetic admixing, presumably there would come a theoretical point where one would have to speak of mixed identity, but such ideas would appear to be highly unlikely

141 James (1997) *op cit*, Note 16

142 Tabashnik, B E (1997) *Proceedings of the National Academy of Sciences USA*, vol 94, pp 3488–3490

143 Emlay, D (1994) 'Compositional Analysis: The Key Component for the Safety Assessment of Flavr Savr™ Tomatoes or Why Would Anyone Want to Feed a Whole Food to Rats', in *The Bio-safety Results of Field Tests of Genetically Modified Plants and Micro-organisms*, pp 209–211, University of California: Oakland, California

144 Tesco (1997) 'Genetic Modification: Questions and Answers', Information Sheet, Tesco Stores plc: Cheshunt

145 Safeway (1996) 'A Guide to Safeway Double Concentrated Tomato Puree produced from Genetically Modified Tomatoes', Information Sheet, Safeway plc: London

146 European Ecumenical Commission for Church and Society (1995) 'The Dominant Economic Model and Sustainable Development – Are they Compatible? A submission to the European Union's Fifth Environmental Programme Review', European Ecumenical Commission for Church and Society: Brussels

147 Burke (1997) *op cit*, Note 131

148 Tesco (1997) *op cit*, Note 144

149 *New Scientist* (1997) 'If it's Safe Then Prove It' and 'Europe Lets in American Supermaize', p 3 and 8, 4 January; and *GenEthics News* (1997) 'Europe Turning Against Ciba Maize' p 4, issue 16, January–March

150 Schoon, N (1997) 'Gene Experts' Maize Mistake', *The Independent*, 11 March

151 *Farmers Weekly* (1998) 27 February

152 Polkinghorne Committee report (1993) *op cit*, Note 133

153 *GenEthics News* (1998) 'Eurochaos over Food Labelling', issue 21 Dec 1997–Jan 1998, p 4

154 Ministry of Agriculture, Fisheries and Food (1995) *op cit*, Note 138

155 Food and Drink Federation (1995) *Food for our Future*, Food and Drink Federation: London

156 Science Museum, (1995) UK National Consensus Conference on Plant Biotechnology; Final Report, London

157 Banner Committee report (1995) *op cit*, Note 45

CHAPTER 7 LETTING OUT THE GENIE

158 Anderson, I (1995) 'Deadly Rabbit Virus out of Control', *New Scientist*, 4 November, p 7

159 Royal Commission on Environmental Pollution (1989) *Thirteenth Report: the Release of Genetically Engineered Organisms into the Environment*, Cmd 720, HMSO: London

160 Under Part VI of the Environmental Protection Act, 1990

161 Tait, J (1990) 'Environmental Risks and the Regulation of Biotechnology' in: Lowe, P, Marsden, T and Whatmore, S (eds) *Technological Change and the Rural Environment*, pp 168–202, David Fulton Publishers: London

162 Levidow, L and Tait, J (1991) 'The Greening of Biotechnology: GMOs as Environment-Friendly Products', *Science and Public Policy*, vol 18, no 5, pp 271–280

163 Sussman, M et al (1988) *The Release of Genetically Engineered Micro-organisms*, Academic Press: London

164 Kornberg, Sir Hans (1988) 'Opening Remarks' in: Sussman, M et al (eds) *The Release of Genetically Engineered Micro-Organisms*, pp 1–5, Academic Press: London

165 Tiedje, J M et al (1989) 'The Planned Introduction of Genetically Engineered Organisms: Ecological Considerations and Recommendations', *Ecology*, vol 70, no 2, pp 298–315

166 Levidow, L and Tait, J (1992) 'Release of Genetically Modified Organisms: Precautionary Legislation', *Project Appraisal*, vol 7, no 2, pp 93–105

167 Jennings D M et al (1975) 'Organophosphorus Poisoning: a Comparative Study of the Toxicity of Carbophenothion to the Canada Goose, the pigeon and the Japanese Quail', *Pesticide Science*, vol 6, pp 245–257

168 Tait, J and Levidow, L (1992) 'Proactive and Reactive Approaches to Risk Regulation: the Case of Biotechnology', *Futures*, April, pp 219–231

169 Stringer, A and Lyons, C H (1974) 'The Effect of Benomyl and Thiophanate Methyl on Earthworm Populations in Apple Orchards', *Pesticide Science*, vol 5, pp 189–196

170 von Moltke, K (1987) *The Vorsorgenprinzip in West German Environmental Policy*, Institute for European Environmental Policy: London.

171 Royal Commission on Environmental Pollution (1989) *op cit*, Note 159

172 Bennett, D, Glasner, P and Travis, D (1986) *The Politics of Uncertainty: Regulating Recombinant DNA Research in Britain*, Routledge and Kegan Paul: London

173 van der Meer, P (1993) 'Potential Long Term Ecological Impact of Genetically Modified Organisms: a Survey of Literature, Guidelines and Legislation', Nature and Environment, no 65, Council of Europe Press: Strasbourg

174 Jasanoff, S (1986) *Risk Management and Political Culture*, pp 59–59, Russell Sage Foundation, New York

175 Her Majesty's Government (1990) *This Common Inheritance: Britain's Environmental Strategy*, Cmd 1200, HMSO: London

176 Council of the European Communities (1989) 'Directive on the Deliberate Release to the Environment of Genetically Modified Organisms', 1 December, 9644/89, Brussels

177 There was a measure of circular argument in this, however. The OECD had exerted a strong influence on both the EC Directive and the UK Regulations, but both the OECD and the EC were strongly influenced by the existing voluntary system of regulation which had developed in the UK. The UK regulations in turn acknowledged their debt to the EC and the OECD as a means of legitimisation of the process of transfer from a voluntary to a legally-based system

178 Department of the Environment (1989) 'Environmental Protection: Proposal for Additional Legislation on the Intentional Release of Genetically Manipulated Organisms, a Consultation Paper', June

179 Royal Commission on Environmental Pollution (1989) *op cit*, Note 159

180 *Hansard* (1990) 6 March, pp 952–3

181 Fleising, U (1989) 'Risk and Culture in Biotechnology', *Tibtech*, March, pp 52–57

182 Lowi, T J (1990) 'Risks and Rights in the History of American Governments', *Daedalus*, Fall, pp 17–40

183 Macrory, R (1983) 'Environmentalism in the Courts' in: O'Riordan, T and Turner, R K (eds) *Progress in Resource Management and Environmental Planning*, vol 4, pp 153–168, Wiley: Chichester

184 Levidow, L and Tait, J (1993) 'Advice on Biotechnology Regulation; the Remit and Composition of Britain's ACRE', *Science and Public Policy*, vol 20, no 3, pp 193–209

185 Girling, R (1990) 'Why Life Will Never Be the Same Again', *Sunday Times Magazine*, 13 May, p 42

186 *Financial Times* (1990) 29 November

187 SAGB (1990) and (1991) 'Community Policy for Biotechnology: Priorities and Actions and Community Policy for Biotechnology: Economic Benefits and European Competitiveness', Brussels

188 Council of the European Communities (1989) *op cit*, Note 176

189 House of Lords Select Committee on Science and Technology (1993) 'Regulation of the United Kingdom Biotechnology Industry and Global Competitiveness', HL Paper 80 and HL Paper 80–1, HMSO: London

190 European Commission Decision (94/730/EC) (1994)

191 European Commission Directive (94/15/EC) (1994)

192 Department of the Environment (1995) 'Guidance to the Genetically Modified Organisms (Deliberate Release) Regulations 1995'

CHAPTER 8 PATENTING LIFE

193 Hatfield, M (1987) Introduction of Moratorium on Animal Patenting. Statement of Mark Hatfield, Congressional Record, US Senate S7268, 28 May, quoted in: Kimbrell, A (1993) *The Human Body Shop*, Harper: San Francisco

194 European Patent Convention, 1978, Article 52(4)

195 European Patent Convention, 1978, Article 53a

196 Mestel, R (1994) 'Cotton Patent Hangs by a Thread', *New Scientist*, 17 December, p 4

197 (1995) 'European Patents: Untapped Potential' *Innovation and Technology Transfer*, vol 2, pp 3–4

198 *Diamond v Chakrabarty*, 447 US 303, 65 L Ed 2d 144, 100 S Ct 2204 (1980)

199 Arthur, C (1993) 'The Onco-Mouse That Didn't Roar', *New Scientist*, 26 June, p 4

200 Nott, R (1993) 'Plants and Animals: Why They Should be Protected by Patent and Variety Rights', *Patent World*, July–August, pp 45–48

201 Nott (1993) *ibid*

202 (1997) Debate transcript: 'What Role, if Any, has Morality in Modern Patent Law?', *Proceedings of the Centennial Conference of the International Association for the Protection of Intellectual Property (AIPPI)*, Vienna, 23 April, AIPPI: Vienna

203 Rothley, W, MEP (1995) 'Why Parliament Must Think Again About Biotechnological Protection', *European Brief*, March/April, pp 60–62

204 Bruce, D M (1997) 'Patenting Human Genes: A Christian View', *Bulletin of Medical Ethics*, January, pp 18–20

205 Reiss, M J (1997) 'Is it Right to Patent DNA?', *Bulletin of Medical Ethics*, January, pp 21–24

206 Crespi, R S (1989) 'Patents in Biotechnology: the Legal Background, Proceedings of International Conference on Patenting Life Forms in Europe', 7–8/2/89, p 7, Brussels

207 Hoyle, R (1995) 'Don't Dismiss Rifkin's Damning of Gene Patents', *Biotechnology*, vol 13, p 643

208 Reiss and Straughan (1996) *op cit*, Note 93

209 Council of Europe (1996) 'Convention for the Protection of Human Rights and Dignity of the Human Being with Regard to the Application of Biology and Medicine', Article 21, November, Strasbourg

210 Crespi (1989) *op cit*, Note 206

211 SmithKline Beecham (1996) *What is the Case for Patenting DNA?*, SmithKline Beecham: London

212 Mooney, P (1989) 'From Cabbages to Kings: Intellectual Property vs Intellectual Integrity' in: *Patenting Life Forms in Europe*, Proceedings of an International Conference at the European Parliament, 7–8 February, p 31

213 d'Silva, J E (1989) 'Patenting of Animals: A Welfare Viewpoint', p 48 in *Patenting Life Forms in Europe*, Proceedings of an International Conference at the European Parliament, 7–8 February, p 48

214 Wilkie, T (1993) 'This Mouse Should Roar', *The Independent*, 13 January

215 Cabinet Office (1992) 'Intellectual Property in the Public Sector Research Base', Office of Science and Technology Report for Rt Hon William Waldegrave

216 Laurie, G T (1995) 'Biotechnology and Intellectual Property: A Marriage of Inconvenience?' in McLean S A M (ed) *Contemporary Issues in Law, Medicine and Ethics*, Dartmouth

217 Durant, J (1994) (ed) 'UK Consensus Conference on Plant Biotechnology, Final Report', Science Museum: London

218 Monti, M (1998) Letter from European Commissioner for Internal Market to the General Secretary of the European Ecumenical Commission for Church and Society, 3 March

CHAPTER 9 GENETIC ENGINEERING AND DEVELOPING COUNTRIES

219 James, C and Persley, G J (1990) in G J Persley (ed) *Agricultural Biotechnology Opportunities for International Development*, CABI: Wallingford
220 UN (1988) 'Transnational Corporations in Biotechnology', United Nations Report
221 Crucible Group (1994) *People, Plants and Patents*, International Development Research Centre, Ottawa
222 Verma, S K (1995) *TRIPS and Plant Variety Protection in Developing Countries*. EIPR (6)
223 Crucible Group (1994) *op cit*, Note 221
224 Crucible Group (1994) *ibid*
225 Crucible Group (1994) *ibid*
226 Mussey (1992) quoted in Crucible Group (1994) *op cit*, Note 221
227 Hindmarsh, R (1991) *The Ecologist*, vol 21, no 5, September–October, pp 196–205
228 Buttel, F H (1995) 'The Global Impacts of Agricultural Biotechnology: a Post-Green Revolution Perspective' in: Mepham, T B, Tucker, G A and Wiseman, J (eds) *Issues in Agricultural Bioethics*, pp 345–360, Nottingham University Press: Nottingham
229 Broerse, J E W and van de Sande, T (1995) 'Technology Transfer or Alternative Technology; Biotechnology and Low-External-Input Agriculture' in: *Issues in Agricultural Bioethics*, pp 345–360
230 Durant (1994) *op cit*, Note 217

CHAPTER 10 THE SOCIAL CONTEXT OF GENETIC ENGINEERING

231 Beck-Gernsheim, E (1995) *The Social Implications of Bioengineering*, Humanities Press: New Jersey
232 Merton, R K (1957) *Social Theory and Social Structure*, Glencoe Free Press
233 Weber M (1918) 'Science as a Vocation' in: Gerth, H and Mills, C W *From Max Weber*, Routledge and Kegan Paul: London
234 Kuhn, T (1962) *The Structure of Scientific Revolutions*, University of Chicago Press: Chicago
235 Rose, H (1994) *Love, Power and Knowledge. Towards a Feminist Transformation of the Sciences*, Blackwell/Polity: Cambridge
236 Brubaker, R (1984) *The Limits of Rationality*, Allen and Unwin: London
237 Weber, M (1978) *Economy and Society*, University of California Press: Berkeley
238 Radford, T (1994) 'Code of Conduct', *Guardian*, 21 July
239 Radford (1994) *ibid*
240 *GenEthics News* (1994) 'Animals Genes and Ethics', issue 2, July/August, p 8

241 *GenEthics News* (1994) 'Coming Soon to a Field Near You', issue 2, July/August, p 12
242 Durant (1994) *op cit*, Note 217
243 Wynne, B (1996) 'May the Sheep Safely Graze? A Reflexive View of the Expert-Lay Knowledge Divide' in: Lash, Szerszynski and Wynne (eds) *Risk, Environment and Modernity. Towards a New Ecology*, pp 44–83, Sage: London
244 Giddens, A (1991) *Modernity and Self-Identity*, Polity: Cambridge
245 Beck (1986) *op cit*, Note 67
246 Beck, U (1995) *Ecological Politics in an Age of Risk*, Blackwell/Polity: Cambridge
247 Lash, S, Szerszynski, B and Wynne, B, (eds) (1996) *Risk, Environment and Modernity. Towards a New Ecology*, Sage: London

APPENDIX 3 FRAMEWORKS FOR MAKING ETHICAL ASSESSMENTS

248 Mepham (1993) *op cit*, Note 82

FURTHER READING

Beck, U (1996) *Risk Society* Sage: London

Beck-Gernsheim, E (1995) *The Social Implications of Bioengineering* Humanities Press: New Jersey

Finchan, J R S and Ravetz, J R (1991) *Genetically Engineered Organisms: Benefits and Risks* Open University Press: Milton Keynes

Kimbrell, A (1993) *The Human Body Shop: The Engineering and Marketing of Life* Harper Collins: London

Kloppenburg, Jr, J R (1991) *First the Seed: The Political Economy of Plant Biotechnology 1492–2000* Cambridge University Press: Cambridge

Lewis, C S (1943) *The Abolition of Man* HarperCollins: London

Lewontin, R.C (1993) *The Doctrine of DNA – Biology as Ideology*, Penguin: London

Maclean, N (ed) *Animals with Novel Genes* Cambridge University Press: Cambridge

Mepham, T B, Tucker, G A and Wiseman, J (eds) (1995) *Issues in Agricultural Bioethics* Nottingham University Press: Nottingham

Ministry of Agriculture, Fisheries and Food (1995) *Report of the Committee to consider the Ethical Implications of Emerging Technologies in the Breeding of Farm Animals* (Banner Committee report) HMSO: London

Northcott, M S (1996) *The Environment and Christian Ethics* Cambridge University Press: Cambridge

Page, R (1996) *God and the Web of Creation* SCM: London

Page, R (1986) 'The Earth is the Lord's – Responsible Land Use in a Religious Perspective' in *While the Earth Endures – a Report on the Theological and Ethical Considerations of Responsible Land Use*, SRT Project, Church of Scotland

Perlas, N (1994) *Overcoming Illusions about Biotechnology* Third World Network: Penang

Reiss, M and Straughan, R (1996) *Improving Nature – The Science and Ethics of Genetic Engineering* Cambridge University Press: Cambridge

Rifkind, J (1991) *Biosphere Politics: A New Consciousness for a New Century* Harper: San Francisco

Schuurman, E (1980) *Technology and the Future*, Wedge Publishing: Toronto

Siva, V (1993) *Monocultures of the Mind: Biodiversity, Biotechnology and the Third World* Third World Network: Penang

Tait, J (1990) 'Environmental Risks and the Regulation of Biotechnology' in: Lowe, P, Marsden, T and Whatmore, S (eds) *Technological Change and the*

Rural Environment David Fulton: London

Thompson, P B (1994) *The Spirit of the Soil: Agriculture and Environmental Ethics* Routledge: London

Wheale, P and McNally, R (1995) *Animal Genetic Engineering: Of Pigs, Oncomice and Men* Pluto Press: London

Wilkie, T (1993) *Perilous Knowledge – the Human Genome Project and its Implications* Penguin: London

Wright, C (1995) *Walking in the Ways of the Lord – the Ethical Authority of the Old Testament* Apollos: Leicester

INDEX